Shell Beach

—

the search for the final theory

Shell Beach - the search for the final theory

Copyright © 2021 Jesper Møller Grimstrup

First English edition 2021
ISBN| 978-87-972868-0-7

Cover design and illustration| **Sten Jauer** at LAND Designlab
www.landdesignlab.com

Cover background photo| Michael Olsen on Unsplash.com

SHELL BEACH

THE SEARCH FOR THE FINAL THEORY

JESPER MØLLER
GRIMSTRUP

The looking itself is a trace
of what we are looking for

— Rumi

To Ria Blanken, Niels Peter Dahl, Kenneth Garrett, Kasper Gevaldig, Kristian Gjerding, Ilyas Khan, Simon Kitson, Jens Christian Haukohl Kristensen, Kasper Køppen, Tero Pulkinen, Anders Led Risager, Christopher Skak, Rolf Sleimann, Carsten Sundwall, and all the supporters of my 2016 crowdfunding campaign.

CONTENTS

PROLOGUE

The reality that we experience has a priori two basic components: there is the world 'out there' and there is us 'in here' who observe it all. This book is a personal account of an attempt by the latter to pin down the former in a closed formula.

The immediate reason for writing this book is that I ran a crowdfunding campaign in the early summer of 2016 to generate funding for my research project with my colleague and friend, the mathematician Johannes Aastrup. One of the perks that I offered to my backers was a book; a book about fundamental physics, about our research, about everything and nothing connected to my work with Johannes.

However, this book has been underway for a much longer time. I have been an active researcher in theoretical physics for more than 20 years and over the past 16 years, Johannes and I have developed a new approach to the problem of finding a unified theory of nature — a theory of everything. This work involves — besides a lot of very technical analysis — considerations of a kind that I believe are of interest to everyone with a curious mind. I have often thought about putting these thoughts down in the format of a book written for a general audience, but I am a mathematical physicist and not a writer and thus I never did so. Until now.

This book is about the pursuit of ideas. It is the idea that a final theory exists — the theory of everything that so many people have sought for so very long — and it is my idea that this final theory will essentially be *empty*. It is the idea that the search for a final theory will come to an end *without* answering our deepest and most urgent questions about our existence.

My pursuit of this idea began in 2002 in Chinese occupied Tibet, where I found the first trace of what Johannes and I now call *Quantum Holonomy Theory* — our candidate for a final theory. It has taken me through several continents and uncountable conferences and workshops in theoretical and mathematical physics, first financed by various universities and research agencies and now by crowdfunding and private sponsors. I sold my flat in Copenhagen to keep this search going and, in this book, I wish to share why I find this work so worthwhile that I have given up almost everything to be able to do it.

Research in theoretical physics is like kayaking a huge waterfall in slow motion. You don't know what you'll encounter when horizontal turns vertical but what you do know is that you cannot do it half-heartedly. Our waterfalls are the ideas we pursue, and like all big waterfalls you cannot scout them thoroughly from the riverbank — the only way to find out is to enter the main current and go over the edge. This book is an account of the largest intellectual waterfall I have ever kayaked, a waterfall that Johannes and I are still tumbling down hoping to find sweet, deep water at the bottom of the fall.

What we have found so far amidst these cascading waters of contemporary theoretical physics is the contour of a theory — *Quantum Holonomy Theory* — which tells us a remarkable tale about the Universe. It is a tale in which the Big Bang was probably not a bang but rather a bounce, black holes have a bottom and the laws of physics appear to emerge from a simple geometrical principle related to empty space. It is a theory with which we are just now becoming acquainted and which will be the main focus in the final chapters of this book.

But the tale I wish to tell is not merely the tale of our theory but also the larger tale of the search itself: the dream of a final theory and those who dream it. *Us.* Why this search and what do we hope to find? Where did it begin and what might we expect to find at its conclusion?

This tale is not only about the science but also the scientists; truth-seeking people like most of us, who wonder about our planet and the infinite void it falls through. It is the human aspect of our search for truth that I wish to write about — it is the *feeling* of adventure, the feeling of going down a huge waterfall for the first time, the feeling of intellectual free-falling.

CHAPTER 1.

THROUGH AKSAI CHIN

Chinese Kashmir,
November 13, 2002

It is cold.

The sun is rising, its first rays carefully sweep across the hilltop above the little Chinese military camp of Mazar and slowly make their descent towards the gravel road where I am sitting.

Chang comes out of his small shack wearing a cap and military outfit and hands me a cup of hot tea with a smile and a few Chinese words. He boiled a chicken last night and offers me some soup for breakfast. His soup, which has the entire chicken in it, feet and head and everything in between, is very spicy and I politely refuse him with a friendly gesture. He speaks Chinese, I don't speak Chinese, and yet we communicate. Chang is from Beijing, where he has a daughter, and he plans to close down his little roadside restaurant within a week and drive to Kashgar before the winter sets in and closes the road. If everything else fails I've thought that we might be able to get a ride out with him.

I have taken a chair outside with me and sit by the road to wait for the few civilian trucks that pass through the military zone of Aksai Chin every day, from the western Tibetan plateau to Kashgar on the old silk road in western China. I am hoping to catch a ride for us, my girlfriend and me.

I have my notebook with me. I want to spend some time today writing down some of the thoughts that have occupied my mind for the past weeks. Thoughts about physics, about quantum mechanics and gravity, thoughts about interactions. Perhaps there is something here that can provide answers to the questions, which I have brought with me on this trip to Tibet.

A week has passed since I lost my passport in a petty little town full of Chinese military, Tibetan pilgrims, and Uighurs in the Ngari region of western Tibet, where the river Indus has its spring and where Tibet becomes China. My girlfriend and I had traveled illegally from Lhasa to Ngari because we wanted to visit the holy mountain of Kailash, a day trip south of town. But when I got off the bus in Ngari I discovered with horror that I no longer had my passport with me. It was nowhere to be found. So instead of visiting a holy mountain that day, we visited the local police station. There we were fined for illegal traveling and got a temporary travel permit and instructions to go to Beijing within 30 days to get a new passport. And finally, a travel ban on Tibet, which meant that we were forced to head west through the military zone of Aksai Chin.

Aksai Chin is the northernmost part of Kashmir, where India and China fought a brief war in 1962 and which China now controls. It is a remote and desolate mountain area of stunning beauty at high altitude, around 5000 meters above sea level. A simple gravel road passes through this territory. From Kashgar it sweeps just north of Pakistan and into western Tibet, about 1000km, the Xinjiang-Tibet-Highway. The road is closed from mid-November due to winter conditions, which means that we might have another week to make it through to Kashgar.

The problem is that it is almost impossible for independent tourists to travel through Aksai Chin. The only possibility is to hire a 4WD with an official guide and driver. But that costs money, lots of money, which we don't have, especially now that I travel without a passport and all our traveler's checks happen to be in my name and can only be cashed by me, holding a passport in my name.

So, we decided to hitchhike. We filled a large bag with supplies and took positions by the road west of Ngari one early morning about a week ago and waited.

The first ride we caught was a truck filled with holy cattle. Three Uighur men from Kashgar had stolen the cattle in India and were driving them across the Himalayas to sell them for slaughter in China. The poor creatures, all bearing the characteristic red spot on their foreheads, had been abruptly deprived of their divine status and were instead treated to altitude sickness and kicks and beatings by the Uighurs. Twice on the trip, the cattle were led out of the truck to graze and to be laughed at by the men who watched them stagger confusedly about in the thin, cold air. A few divine creatures died of altitude sickness on the trip and they were probably the lucky ones.

The trip with the Uighur cattle thieves took us about 400km into Aksai Chin, where they dumped us in a small mountain village in the middle of the night. I think they were afraid that they couldn't get us through some of the military checkpoints. We did pass one checkpoint with them: they hid us behind the seats in the truck while they talked to the Chinese soldiers and gave them cigarettes. That was all well, but I think they knew that it wouldn't be as easy to smuggle us through the checkpoints closer to Kashgar and so they left us. It was bitter cold, but we found a little inn type of thing where we spent the night. After that, we had one hazardous ride with yet another Uighur cattle transport — this time with Yak oxen and a drugged driver — and then we ended up here in Mazar with Chang Zu, our newfound Chinese friend, who is just now attempting to start the engine in his jeep to get electricity for his makeshift kitchen.

I am flicking through my notebook while I wait. I know almost nothing about China. I have no guidebook and no map. I've got a basic sketch in my book, where I have placed the most important cities: Kashgar, Urumqi, Beijing, and connected them with lines symbolizing railroad connections. I also have a collection of the 10 most important phrases in Chinese: "Where is the train to ...,", "We are looking for a hotel", "Where do I find a toilet?", etc. My phrases are written first in English and then in Chinese, so all I have to do is show the page to people I meet and hope for a comprehensible reply.

Two months have passed since we left Vienna. I defended my Ph.D. in August at the technical university of Vienna and on September 11, 2002, we left for Tibet. We have almost two months until I begin my job as a postdoc at the University of Iceland in Reykjavik. If we can catch a ride out of this dump and escape the winter, that is.

I watch the mountains in front of me for a while. They form a massive wall to the southwest. It must be Pakistan out there. K2 should be somewhere in that direction, about 150km away. And Nanga Parbat too. And Broad Peak. I have always wanted to travel through the

Karakoram in Pakistan and now we're so close. If, instead of turning north towards Kashgar we continued west, then we would reach the Pakistani border within a day or so. But winter is coming, and I don't have a passport.

I find an empty page in my notebook and start writing my thoughts, all the stuff I have been pondering in the past weeks while hiking in the Himalayas.

When I was finishing my Ph.D. in the early summer I began to think very carefully about the foundation of my thesis. My thesis is an analysis of a simple idea on how to extend quantum mechanics to include also time and space as quantum operators. The idea goes back to Wolfgang Pauli, one of the fathers of quantum mechanics, and has become popular in the past years since it was shown to be connected to a certain string theory limit.

What I was contemplating, while I sat in Vienna and wrote the final pages of my thesis, was a theoretical consideration that eventually made me question the entire foundation of my thesis. Technically everything was kosher, it was all solid, I was certain of that, but my entire work — and with it, a considerable body of scientific work published through the past several years — was based on assumptions, which I suddenly no longer believed could be true. In fact, I'd become convinced that they were wrong.

I brought all those thoughts with me to Tibet. On a long hike through some very remote and incredibly beautiful mountains on the northeastern side of Everest, I kept turning them over and over in my mind. It was like an obsession. I walked with a ski pole in one hand and a rope to the yak ox that carried our supplies in the other hand,

and wondered what mathematical assumptions could replace those, which I had rejected.

The problem that I was thinking about is the reconciliation of quantum mechanics and Einstein's theory of relativity and it is about how to interpret and explain the mathematical structure of the Standard Model of particle physics — the mathematical theory, which has been so incredibly successful in explaining our reality on a subatomic scale. The problem is finding a final theory that explains all the physical phenomena that we know.

And those thoughts had begun to change and take shape on that long hike. A sensation in my stomach had arrived, a good feeling. The feeling of an idea, a good idea. And it is this feeling, which fills me now as I sit by this Chinese roadside drinking Chang Zu's steaming tea. I'm going to write down this idea in my notebook.

CHAPTER 2.

THE SEARCH FOR A FINAL THEORY

*"But where shall wisdom be found?
And where is the place for understanding?"*

— Book of Job

*"Vanity of vanities", saith the Preacher,
"vanity of vanities; all is vanity".*

— The Ecclesiastes

Imagine the existence of a single mathematical principle that explains the entire physical reality and that cannot itself be reduced to other, deeper principles. A principle that explains all of it. All the known laws of physics, the Universe, the cup of black tea I drink while I write this, the grass I sit on, the air I breathe.

Imagine if the tower of explanations that modern science offers — where biology stands on chemistry, which again stands on atomic physics, nuclear physics, and particle physics, and where cosmology provides the overall wrapping — imagine if that tower of scientific theories, the result of centuries of probing ever deeper into the fabric of reality, came to an end and we found a final theory.

Such a theory would form a foundation for everything we know about the physical reality. It would be final in the sense that it clearly could not be explained by yet another, deeper layer in the way that, for instance, chemistry can be understood in terms of atomic physics. It would be the completion of reductionism.

Can you feel it? The rush, the thrill?

Does such a theory really exist? Is it waiting for us to find it, somewhere out there in the land of mathematics, in the realm of intellectual, logical reasoning? And if so, what could it possibly look like? What would it tell us about reality? What would it tell us about us?

The search for such a theory is the central theme in modern theoretical physics. Thousands of scientists through generations have dedicated their professional lives to this search. Some lost their faith underway and stopped believing that such a theory exists; a few came to believe that they had already found it; and many others remain hopeful and silently pray that they will live just long enough to see the day when this holy grail of science is found.

The idea that there exists a fundamental principle that explains the entire world goes back to ancient Greece and the first great civilizations in the middle east. The Greeks were the first in human history to see reality as a question that requires an answer. They were the first to make the crucial shift from merely philosophizing about reality to base their investigations on actual empirical data and mathematics. They began to ask reality questions and they wanted answers. Thereby they planted the seed for that intellectual discipline, which we today know as the scientific method and the scientific project, and which has swept through the centuries like a wave, from generation to generation and culture to culture.

To make theories and to hold them up against empirical evidence.

But what makes theoretical physicists today believe that a final theory exists? What have we learned since ancient Greece that makes us think that we can find it?

To answer this question we first need to look at where theoretical physics stands today. This will occupy several chapters of this book, in which we will first visit quantum mechanics, Einstein's theory of relativity and the standard model of particle physics, and thereafter the more contemporary subjects of non-commutative geometry and loop quantum gravity.

But before we begin this tour and all the detours and odd adventures that come with it, let me first state the obvious: that the possible existence of a final theory is wildly debated and that we are nowhere near a consensus. In my work with the mathematician Johannes Aastrup, we have proposed our candidate for a final theory, which I will discuss in some of the final chapters in this book, but we are certainly not in a position to claim that we have a definite answer. Nevertheless, there are observations and arguments that suggest that a final theory does exist and that one day our search could come to an end.

The Nobel laureate Steven Weinberg writes in his book 'Dreams of a Final Theory' about the convergence of arrows of explanations:

> "... *Think of the space of scientific principles as being filled with arrows, pointing towards each principle and away from the others by which it is explained. These arrows of explanations have already revealed a remarkable pattern: they do not form separate disconnected clumps, representing independent sciences, and they do not wander aimlessly — rather they are all connected, and if followed backward they all seem to flow from a common starting point. This starting point, to which all explanations may be traced, is what I mean by a final theory.*"

Weinberg believes that

> "... *it is very difficult to conceive of a regression of more and more fundamental theories becoming steadily simpler and more unified, without the arrows of explanation having to converge somewhere.*"

As Weinberg writes, this convergence of explanations is seen throughout the natural sciences. Take for instance the periodic system that maps the chemical elements and their basic properties. Viewed by itself, the periodic system displays a high degree of complexity, but it can be explained by atomic physics and our knowledge of electron orbits. Or take particle physics, where a complex zoo of different particles with different properties is well understood in terms of the standard model of particle physics, a relatively simple mathematical theory.

One of my favorite examples of the convergence of explanations comes from electromagnetism, an example that also demonstrates how theories, once discovered, often carry the seeds for the next steps in the scientific evolution.

Before Ørsted discovered the connection between electricity and magnetism these phenomena were understood simply as different physical occurrences. Like apples and polar bears. Even with the formulation of Maxwell's equations, which provides the precise relation between the electric and magnetic fields, we are still dealing with different phenomena. There is the electric field and there is the magnetic field. This is much like Einstein's field equations in general relativity, where you have the relationship between the geometry of space and time and the energy content of matter. Again, different phenomena brought together in a set of equations.

If, however, Maxwell's equations are reformulated in terms of a new type of field, known as a gauge field, that combines both the electric and the magnetic fields into one entity, then these equations are suddenly seen to have a very particular mathematical structure, which points towards — and opens the doors to — several different theories in modern physics.

First, there is Einstein's special theory of relativity (and with it, indirectly, also the full theory of general relativity), whose structure lies dormant in Maxwell's equations as a hidden message only to be read by those clever enough to decode the mathematics. Second, there are all the generalizations known as non-Abelian gauge theories, which we today know describe the strong and electroweak nuclear forces (and gravity too) and which form the backbone of the standard model of particle physics. And third, hidden deep within electromagnetism one even finds a door that leads us to the quantum world, quantum mechanics, and all its wonders.

Physics is full of such examples, where those who study the mathematical structure of the fundamental theories are left wondering where all this structure originates from and what it all points towards — what it means. The arrows of explanations do indeed appear to converge and their point of convergence seems to lie just beyond the horizon.

✡

There exists a simple argument based on quantum mechanics and Einstein's general theory of relativity, which suggests that the process of scientific reduction must end somewhere.

This argument begins with one of the key formulas in quantum mechanics, namely the Heisenberg uncertainty relation. This relation states that the combined uncertainty of the measurement of the position and of the momentum (mass times velocity) of a particle must always be greater than a certain number known as Planck's constant (in chapter 4 I will discuss quantum mechanics in more detail).

Thus, if we want to measure the position of an object with great precision then the uncertainty in momentum will grow proportional to this precision, according to quantum mechanics. This means that the smaller a region in space that we wish to probe, the higher the energy transfer to that region will be. However, Einstein's theory of general relativity tells us that this energy will cause space and time to curve and that this curvature at some point — if we keep measuring ever shorter scales — will become so large that not even light can escape (in chapter 6 we'll take a closer look at Einstein's theory). That is, the probe that we use for our measurement, will create a black hole.

So, distances smaller than a certain scale — which is known as the Planck scale — are operationally meaningless. No matter what probe we use for our measurement we cannot get a signal back because of the interaction between the probe and space itself. Distances shorter than the Planck scale cannot be measured.

But each scientific reduction — from biology to chemistry to atomic physics etc. — involves a jump to a smaller scale. That is what reductionism is all about, we climb down the ladder of ever shorter scales. So the possible existence of a smallest physical scale seems to offer a very strong hint that this ladder of scientific reductions must end somewhere. It cannot continue indefinitely.

This argument is not a proof but seen from the distance it certainly could be mistaken for one. There could be a long way to go, though. The Planck scale is incredibly small, 10^{-33} cm, which means that there is room for a very large number of layers in this onion of a Universe that we are peeling.

So, we find ourselves in the fascinating situation where we have good reason to believe that a final theory exists, while finding it almost impossible to imagine what such a theory could possibly look like. What theory can be immune to the question "Why is it this way and not another way?" What mathematical structure could possibly stand up to the scientific process of reductionism? Is such a theory even conceivable?

This is where we start, our journey begins with one of the most interesting intellectual challenges imaginable.

✡

But before we throw ourselves into the fight, let us first return to the ancient Greeks. I would like to understand why they set all this in motion – the scientific project and the scientific method. Why did they do it? What did they hope to accomplish? And later, with Newton, who discovered classical mechanics and gave us a real sense of what a final principle might look like, what did he expect to find?

And what do we, scientists today, hope to find?

If we truly believe that a final theory exists and that we may be able to find it, then I think it is worthwhile to consider whether such a theory could possibly meet the expectations of those, who commenced this search — and of those of us, who aspire to complete it.

Clearly, the ancient Greeks had none of the scientific insights we have today nor did they have centuries of scientific tradition to stand on. What they did stand on were the mythologies of the ancient civilizations.

In the ancient civilizations, priests had observed the movements of the stars and planets and realized that there was a system to what they saw, that there were reoccurring patterns. With these observations, the idea of a cosmic order, which holds everything in place and which can be formulated mathematically, was born. This cosmic order did not only apply to the heavenly bodies they observed and whose movements they were able to predict by application of mathematics, the cosmic order also applied to society.

The people of these first great civilizations had gone from being generalists to specialists, from being amateurs of all trades to professionals. That transformation had presented society with the problem of providing its individuals — the governing class, the priesthood, the traders, the craftsmen, the farmers, etc. — the experience of being members of a single organism in order to prevent a disintegration of society. Thus, the idea of a cosmic order was not only an interesting observation of philosophical value, it was an important idea that was needed to preserve the increasingly complex social order of society. It was useful and perhaps even necessary to have a mythology that matched the development of society and the idea of a cosmic order provided the foundation of such a mythology.

It is important to note that this cosmic order is an impersonal force, it is an unstoppable process, which from our vantage point resembles a law of Nature. I imagine that this idea must have been part of the platform that the ancient Greeks stood on.

So science grew out of ancient mythologies. But is science itself a kind of mythology?

Any mythology must be able to establish a relationship between the individual and the cosmos. It must offer an answer to the question "why am I here?" This is the key function of any living mythology, and it must do so in a meaningful way that leaves us filled with purpose and peace.

I imagine that when the ancient Greeks took the first leap away from the mythology they stood on and jumped into the unknown, they must have expected to find a new mythology. Something different, of course, something more rational, more in line with the way their rational minds perceived the world, but ultimately a new mythology.

The point I wish to make is that the scientific project is not merely about understanding reality, it is also about understanding our role in this reality and establishing a relationship between us and the Universe. It is important to realize that at the core of the scientific project lies also an existential and ultimately religious question. It really is about mythology. What is our role in this grand theatre of things?

But science is not a new kind of mythology. We are the first civilization that does not have a living mythology. Science has replaced our ancient mythologies but has so far not been able to function as one.

I am convinced that the existential vacuum caused by the absence of a functional mythology is part of the explanation of why we today continue the scientific tradition with such unrelenting force. We have lost something in our search for scientific truth and that something is our place in the grand order of things. We may not admit it, we may even deny it, but I am convinced that the persistence of our continued scientific quest — or conquest — must partly be understood in terms of this absence of a functional mythology.

Could a final theory give that back? Could it make room for a new, modern mythology that does not leave us cold the way science does today? I don't know the answer to this question but I would like to spend some pages of this book to discuss it. Because I believe that this is the truly interesting question to ask. What will the discovery of a final theory mean for us, for our lives, for our way of seeing the world? Where will it leave us? And in which direction will it send this little ship of ours called mankind?

CHAPTER 3

THE SMELL OF AN IDEA

August 6, 2016

I've found the little blue notebook that I wrote in during my travels in Tibet almost 14 years ago. It's worn, the cover is dirty, and some of the pages are torn. I'm flicking through the pages that I wrote over the four days I sat by the road in Mazar waiting for a ride to Kashgar. A few photos fall out of the book: there is one of the Chinese military convoy that finally gave us a ride out, and another one of my ex-girlfriend standing by a food stand in Kashgar. I put them back and find the page where I describe the idea that I got on the hike in the Himalayas. This is what I'm curious about, I want to see what I wrote about the idea.

As I start reading I soon find myself back in Tibet. In the sentences, I rediscover the feelings I had back then. There is excitement and joy, and hope too, lots of hope. There is the adventurous spirit of a young soon-to-be postdoc, impatience and naivety. I recognize him, the man who wrote this, it's a long time ago but I can still feel him. I remember what it was like that day, the cold, the slight fever I had. It's all there in my old notebook, all the feelings, the sensations, the smells, the impressions, written between my words.

But I cannot find the idea. What I read seems like a naive and helpless attempt to write something down, which barely makes sense. There are several sentences with "my idea ..." but no matter how carefully and with how much goodwill I read through my writings I cannot find anything that looks like a real idea.

It isn't there.

I close the book and sit for a while, wondering. I'm sitting under the large birch tree in the garden; I lean back and look up into the branches that slowly sway in the wind. When I close my eyes it sounds like a thousand faint voices just beneath the threshold of intelligibility, like an ocean of lost meanings.

I look at the notebook again. I think that what I found in Tibet back in 2002 was the feeling of an idea. I found the *sensation*. The smell. The idea itself must have come later.

CHAPTER 4

THE QUANTUM

"I daresay you haven't had much practice,"
said the Queen. "When I was your age, I
always did it for half-an-hour a day. Why,
sometimes I've believed as many as six
impossible things before breakfast."

— Lewis Carroll,
Through the Looking-Glass

The challenge you face when you want to explain quantum theory without resorting to mathematics is that although quantum mechanics has been known for almost a century we have not yet reached a consensus on how to interpret this amazing and surprising theory.

We know the mathematics of quantum mechanics very, very well. There are still open questions concerning mathematical aspects — there always are in science — but overall we have a very solid understanding of what is going on. We know this and we have known it for a long time.

And yet, when we try to explain the mathematics to each other and to people outside of physics, it gets hard. I recall a conference I attended a few years ago at the mathematical research center Oberwolfach in southern Germany. It had been a week full of technical talks about mathematical physics but then, towards the end of the conference, there was a talk by a Dutch physicist about the interpretation of quantum mechanics. The talk was brilliant and the reaction of the audience remarkable: what had been a relatively quiet, calm, and highly professional crowd of specialists for the past week suddenly erupted in an emotional debate about philosophy and what it means to be 'real'.

What was going on is this: when we try to bridge the gap between the mathematics of quantum mechanics and the language of our everyday lives, we are almost immediately faced with what appears to be absurdities. The quantum world seems to completely defy common sense. What we encounter there is so different from what we experience elsewhere that we feel forced to question language itself and its ability to encompass this realm. But since the quantum world *is* the world, this leaves us in an awkward position. This is what makes it all so incredibly interesting.

There simply is no consensus here. Last I checked, Wikipedia listed almost 20 *different* interpretations of quantum mechanics. The confusion is that great.

In the following I will try to explain in non-technical terms what a quantum theory actually is — as a *mathematical* object. I will avoid the issue of interpretation as best I can because the challenges we are faced with when working on the problem of finding a fundamental theory are foremost technical. The question of interpretation is, in my opinion, secondary, and as long as we have not yet found the final theory we cannot know if what we are trying to interpret is in fact a distorted and incomplete image. It is like approaching a distant ship at high sea: at first, you see only the top of the mast and if you didn't know better you could be tempted to conclude that what you see is the antenna of a crashed UFO calling an intergalactic tow service for assistance.

✡

The process of measurement plays a key role in quantum mechanics.

When you measure something — the location of some object for instance — you need to interact physically with whatever it is you wish to measure. This interaction can, for example, be in the form of light. You shine a beam of light into a dark room, the light reflects off the old suitcase full of pictures from your childhood on the farm out west — the lake with the trout that your dad taught you to fish; the grassy fields where you chased your first dog around; the haystack, where you one summer discovered a batch of newborn kittens — and some of that light reflects back to your eye, where nerve cells communicate with your brain to form a mental image that tells you that the suitcase you are looking for is standing just a few meters to the right.

This beam of light carries energy and some of this energy is transferred to the suitcase during this interaction. This means that your measurement will change the state of what you are measuring ever so slightly: you add some momentum, some energy. Of course, you can always minimize this transfer of energy, and in the case of a suitcase it hardly matters: you'll never experience that the light beam from your flashlight sends the stuff stored in your attic flying around the room. This interaction is extremely tiny.

However, when we enter the atomic and subatomic world, this interaction begins to play an important role. In fact, it will eventually take over the entire act.

The key new element in quantum mechanics is the discovery that there exists a minimal interaction. Nature has a fixed number, which tells us that there is a certain amount of interaction that we can never avoid. No matter how careful we are, no matter how ingeniously we design our experiment, there will always be a minimal interaction between our measurement and the object that we measure. This minimal interaction, given by a constant called the Planck constant, is unavoidable and will *always* change the state of the object.

And this changes everything.

One of the first consequences of this unavoidable smallest interaction is a radical shift from talking about physical quantities to talking about the *measurement* of physical quantities.

This is a shift from *numbers* to the *process* of obtaining these numbers.

In classical physics — and we call everything prior to quantum physics 'classical' — the position of a particle is given by a set of coordinates, i.e. *numbers*. The velocity of a particle is given by a *number*, say 50 kilometers per hour. The position of your garden gnome is given by a *number*, say 33 feet northwest of the apple tree. The kinetic energy of a piano falling from the third floor in your house is given by a *number*, about 20 kilojoules. Everything we know about a classical physical system comes in the form of *numbers*: energy, angular momentum, momentum, position, etc.

In the quantum world, we deal instead with the *measurement* of these quantities. We have the *measurement* of position, the *measurement* of momentum, the *measurement* of energy, etc.

The point is that the discovery that Nature has an unavoidable smallest interaction given by the Planck constant forces us to completely change the mathematical way of describing physical systems. We are forced to abandon a type of mathematics where, for instance, the position of a particle is given by a number and instead adopt a new type of mathematics, where the position of a particle is treated in terms of a *process*.

What is the difference between a number and a process? Well, there can be a huge difference. First of all, numbers do not care in which order we write them. The mathematical term for this is to say that they *commute*. We know that two times three equals six, which also equals three times two. It does not matter which number we write first and which one we write last.

This is no longer the case when it comes to a process. Processes do not always commute. Take for instance the process of climbing down a rope from your lover's apartment on the tenth floor because her husband came home unexpectedly. And the second process of tying that rope around the large fridge in your lover's kitchen. It truly matters which process you execute first. If you climb down the rope *before* tying the knot, you die; if you tie the knot first, you live (for a while, anyway).

The order of these processes is critical, they do *not* commute, and the reason for this is that such processes *change* the system.

This is the key point: processes change the system upon which they act.

The mathematical word we use for such a process is an *operator*. For instance, the operator that corresponds to the position of a particle or the operator that corresponds to its momentum.

Now, the example with the rope is very obvious, everyone understands that you should not climb down a loose rope, not even with a hostile husband in the room. But in quantum mechanics, we encounter processes that do not commute, which are less obvious.

The most important example of such operators is given by the operators that correspond to the position and momentum of a particle. The so-called *canonical commutation relation* describes how these operators fail to commute. And this failure to commute is in fact precisely the mechanism that imposes a minimal interaction on the formalism, i.e., the Planck constant.

"But tell me, why is there a minimal interaction given by the Planck constant?" you might ask.

Well, we don't know. That's the answer. We see the '*how*', but not the '*why*'. Perhaps the answer is simply that

Nature chose it to be this way. Or perhaps we will someday find an answer that nobody has yet thought of.

But if you insist, I might be able to do a little better than that. In my work with Johannes, we may have something that might offer an explanation. Sort of, anyway. But we will get to that later, now is not the time.

✡

A quantum theory consists of two key ingredients: an *algebra* of operators and a mathematical stage, where these operators act, which is called a *Hilbert space*.

So, an operator is a mathematical process that acts on something according to a formal rule. For instance, it may simply be the multiplication of a variable or it may be a differentiation, which is the mathematical process of computing the *rate of change* of some function.

And a set of operators form what is called an algebra, which can be thought of as a general mathematical system, where you can add and multiply different elements. Ordinary numbers also form an algebra: if you add or multiply two numbers you get a new number, and likewise, you can add or multiply different operators to build new ones — it's like playing with LEGOs.

But operators cannot stand alone when we want to form a physical theory. We have replaced physical quantities such as position, momentum, and energy with operators — mathematical processes — but we still need somehow to get *numbers* out in the end. This is what the Hilbert space does.

The Hilbert space is the place, where the algebra of operators *lives*. An operator is a mathematical process and this process has to *act* on something. This something is a *state* in the Hilbert space[1].

Think of the rope again: here the operators are the *actions* of climbing down the rope and of tying the knot. The rope itself is a part of the Hilbert space. It is the object to which we apply these operations; the state of the system.

The Hilbert space is then the collection of *all possible states* that our physical system can be in (tied rope, untied rope, burnt rope, painted rope, tangled rope, whatever rope). And given a state in a Hilbert space, we can compute the likelihood of an outcome of a measurement of some physical quantity — a number! — for instance the state of the rope or — in the case of ordinary quantum mechanics — the position of a particle.

So the algebra and the Hilbert space are like two halves of a complete whole. They need each other to be of real consequence. If you think of an operator as an *actor* in a play, then the algebra is the full ensemble for the play, say Shakespeare's Hamlet (where we have Hamlet, the king, the queen, Horatio, etc.). But the play is nothing unless you have a stage to play it on. It can be the courtyard in the Helsingør Castle north of Copenhagen (where I watched Jude Law play Hamlet a few years ago), it can be on Broadway or anywhere you like, but you need a stage. And once the operators have that, a Hilbert space, then they can give us what we came for, the *show*, which is predictions for our measurements, i.e. *numbers*.

---

[1] A state in a Hilbert space is also called a *wave function*.

So this is the setup I would like you to remember: a quantum theory consists of two parts, the **algebra** of operators and the **Hilbert space**. This is what you need to take with you into the following chapters of this book.

✡

But let me now tell you an interesting, technical detail: not all algebras of operators have a Hilbert space to live in. For some algebras, it is simply impossible to find a matching Hilbert space. Such algebras, as interesting as they might be, cannot form a quantum theory.

So the algebra and the Hilbert space are two halves of a whole, but they are not equal partners: the algebra is primary, its Hilbert space secondary. It starts with the algebra; everything starts with the algebra.

✡

But how do you know what operators to choose?

Different choices of operators correspond to different quantum theories. You can have the quantum theory of a single particle in empty space, you can have the quantum theory of two particles that interact with an external field and you can have a quantum theory of electromagnetism, called quantum electrodynamics or QED. You can cook up an infinite number of different quantum theories.

But of course, we want to know how to find a quantum theory that corresponds to a given physical system. To do so we have a cooking recipe known as *quantization*.

Quantization is a mathematical formalism, which tells us how we get a quantum theory from a classical one by interchanging quantities such as position and momentum variables (which are, I say it again, essentially, *numbers*) with the corresponding operators, and how to find a Hilbert space for them to live in.

It is, however, not a one-way street to go from classical to quantum via quantization. There are ambiguities and unknowns depending on which type of system you want to quantize; there are situations, where we know exactly what to do and there are situations where we are completely in the dark. Quantization is a cooking recipe, but like any recipe, there is room for variation — you can play with other ingredients, like leave out the carrots and add some cabbage instead. And not all such variations lead to an eatable outcome.

Consider for instance electromagnetism. In this case, we know what we need to do to obtain the quantum theory — *quantum electrodynamics* or QED — and the theory has been tested experimentally to an extraordinary high precision. On the other hand, consider Einstein's theory of general relativity. Here we do *not* know what to do in order to obtain a quantum theory — all the known methods of quantization have failed.

✡

Let us pause for a moment and consider one of the most important examples of a quantum system, the hydrogen atom.

As a classical system, the hydrogen atom is a disaster. One negatively charged electron orbiting a positively charged nucleus is a recipe for calamity: the orbiting electron will radiate off electromagnetic radiation and as a consequence immediately crash into the nucleus. The hydrogen atom simply cannot exist. Classically.

But what happens when we quantize it? Well, that's what the fathers of quantum mechanics tried. They took the equations of the classical system — the energy of an orbiting electron — and replaced the classical position and momentum variables with operators.

And when they did this they discovered a set of equations, which they were able to solve. And what did they find? They found a discrete system of electron orbits, which secured the stability of the hydrogen atom and from which they obtained the emission spectrum that they already knew from experiments. It all matched perfectly. It all made sense.

Mathematically, at least. The math made perfect sense.

Let us go back to the Hilbert space. As I said, the Hilbert space consists of all the possible states that the physical system can be in, and when an operator acts on a state it may change it to a different state in the Hilbert space. And because operators can be added — this is how an algebra works, you can add and multiply its elements — then we must also be able to add states. This is what is called a *superposition*.

The existence of superpositions implies that we can have states that do not correspond to a *single classical* configuration of the system but instead have the form of a *sum* of completely different configurations. If we consider an electron, then we will have states that are the sum of the electron being right here and the electron being on the moon.

Or consider the example with the rope: in a quantum theory of *'ropes in high buildings with angry husbands'*, there will be states, which are the sum of a tied rope and an untied rope.

This sounds crazy and it *is* crazy but that is simply the way quantum mechanics *is*. It is a direct consequence of the existence of a smallest interaction.

However, we never come across objects that are both here and on the moon simultaneously, or ropes that are both tied and untied at the same time. The reason is that the states in the Hilbert space effectively encode statistical information about the outcomes of measurements. So when we measure the electron it will be *either* here or on the moon. When we climb out the window — which in this example constitutes a measurement — then we'll *either* live or die.

Now, since the Hilbert space involves all kinds of crazy states, most of which are far away from the classical world, it raises the question of whether the Hilbert space also includes states, which *are* close to the classical world. This is the question of finding a classical limit.

It is crucial for a quantum theory to have these near-classical states, which provide a limit, where the reality of our everyday world emerges. A quantum theory must describe *both* the quantum world that we encounter in experiments *and* the classical world that we live in and where quantum effects are never seen.

✡

Let's now take a moment to try to understand why the interpretation of quantum mechanics has caused so many people headaches for so many years.

In the quantum world, we have thrown away the ordinary — the classical — description of the world in terms of numbers and have replaced it with a mathematical framework based on operators. This means that strictly speaking, we can no longer talk about the position of a particle independently of the measurement hereof. In fact, the position of a particle by itself is <u>not</u> a part of our mathematical description of the system[2]. The only thing we *can* talk about is the *measurement* of the position of a particle.

Yes, okay, one might add, but surely the particle that we are measuring is *out there* somewhere and has a precise position also when we do not measure it?

---

[2] of course, we have the state in the Hilbert space, but that does not give us the position of the particle but rather a statistical likelihood for a given outcome of a measurement.

But how would you know this? The only way to determine whether a particle has a position somewhere is to perform a measurement.

The point is that the physical reality of, say, a particle independently of its measurement is no longer a part of our description of reality. It is not a part of our theory and in a very real sense does not seem to make sense. All we know about is the measurement — in a broad meaning of the word — and the idealized description of the physical system in terms of operators.

This is why the mathematical physicists at the Oberwolfach meeting that I attended a few years ago started shouting at each other when they discussed the interpretation of quantum mechanics — almost a century after its discovery. Quantum mechanics puts into question what is real and if there is one topic that most people have strong opinions about, it's reality.

I vividly remember one of my first lectures in quantum mechanics in my second year at university, when our professor explained one of the most extraordinary experiments in quantum mechanics, the double-slit experiment. Here you fire a particle against a wall with two small slits and measure where the particle hits on a screen behind this wall. First, you see that the particle makes a pattern on the screen indicating that it went through *both* slits and interfered with itself. Thus: a wave. But when you then go on to measure whether the particle travels through one of the two slits or not — you add a second measuring device there — then the measurement on the back screen changes radically: now the particle behaves like a particle, not a wave. No interference pattern.

A wave. A particle. What is it? Where is it? *Is* it?

This lecture was a complete shock to me. It hit me like a 500-tonne locomotive — I recall walking home that day thinking about the experiment and feeling the ground disappear beneath my feet. It demolished my view of the world, and in its place I experienced a deep sense of intellectual dizziness, a dizziness that stayed with me for years.

What is going on here is the dismantling of a worldview. It is the worldview of Newton, where we deal with separate, independent *objects* located in *space* and with *time* as a separate 'tick-tock' of, well, *time going by*. This worldview is central to our experience of the world we live in, it permeates our language and, as Kant suggested, it seems to be a necessary precondition for our very perception of reality. But with quantum mechanics and Einstein's relativity theory, which we shall discuss next, this Newtonian worldview has begun to crumble — and it has not yet been replaced with something new.

In quantum mechanics, we are forced to reconsider the very concept of an object, since it is no longer clear what an object is independent of its measurement. This immediately casts doubt on whether the physical reality, the out there, is, in fact, really out there at all. And with that, we tumble into the abyss of interpretations of quantum mechanics.

What lies at the heart of the problem, I believe, is the word 'exist'. What does this word actually mean? Well, for a 'something' to exist it needs to exist 'somewhere' and it needs to do so in a definite, independent sense of the word. But this is problematic because in quantum mechanics the 'somewhere' is a quantum operator — the position operator — so the assignment of a position can *only* happen in connection with its measurement. This implies that the nomenclature 'exists' cannot be assigned independently of measurement, which defies the original meaning of the word. Most people would balk at the idea that something only exists when measured.

Some counter this by declaring that the state (also called the wave function) exists. That may be so, but in my opinion, this is a huge deflation of the word 'exist' that just shows the uselessness of the word and the quagmire that quantum mechanical interpretations really are. If the state itself exists then so do all the infinite (not even countable) different possible realities it represents, which leads to the so-called many world interpretation, a truly absurd idea.

All these difficulties arise, in my opinion, from clinging to the Newtonian worldview and the idea of objects that exist independently of everything else. This is a mental construction, a metaphysical framework, that we use when we think about reality, but which has, in effect, no physical implications nor justifications by itself. We can just let go of it.

What would happen if we stopped using the word 'exist' in an ontological sense? (an everyday use of the word is certainly useful). From the point of view of science, nothing at all. From the point of view of philosophy and metaphysics: everything. Because it's a metaphysical word, it is part of a mental construction that we can — I say it again — just let go of. It makes no difference.

The word 'exist' has no practical implications at all. What matters is what we measure — understood in a broad sense.

The point here is that if we stick to mathematics we are doing fine. It is when we switch back to our Newtonian worldview and ask questions based on it that we run into trouble. A question like *"where is the electron when we do not measure it"* is based in this worldview (it assumes that the electron is an object that has an independent existence somewhere in space. It assumes that the electron is a *thing*). These questions are simply meaningless, but because we have not yet let go of this mental construction, we find this difficult to accept.

Again: what difference does it make where the electron is between measurements? (or even if it *is* anywhere, again: we find the word 'exist' on our tongue). It makes no difference because to determine its position we would need to perform a measurement.

The problem is not reality, it's the questions that we ask and the worldview that forms these questions. We ask physics to provide us with metaphysical answers. It cannot.

I suspect that the term *'interpretation of quantum mechanics'* in reality translates into *'how to understand quantum mechanics in terms of a Newtonian world view'*. And if that is true, then it is an impossible undertaking.

With Einstein's theory of relativity, which we shall soon discuss, the second part of the Newtonian worldview starts to falter as space and time are no longer merely a passive container for *stuff,* but are in fact shaped by it. With Einstein, we are forced to let go of the idea of space and time as an independent frame of reference, and instead, we learn that the very presence of matter in space and time changes it — according to the mass and momentum it carries.

The Newtonian worldview is simply wrong. It is falling apart and the sooner we get out of the rubble the better.

I am convinced that this general theme of letting go of the Newtonian worldview will continue as we dig deeper into the fabric of reality. Quantum mechanics and Einstein's theory of relativity have given it a serious beating and it seems highly plausible that combining them into a theory of quantum gravity — if such a theory exists! — in which the process of measurement of space and time itself can be described in the language of quantum theory, could destroy it completely. The question, "does space and time exist when we do not measure it", seems to completely escape rational thought. How will it be possible to think about such a theory outside the realm of mathematics? If space and time can exist in superpositions of *different geometries*, if we are talking not only about a superposition of an object *in* space but of space *itself*, then it seems that we have finally reached the point where our Newtonian notion of objects in space and time has been completely dismantled.

✡

A few years ago I taught some classes in quantum mechanics at the University of Iceland. I recall one day when a student —a very bright and quiet guy who usually kept to himself — came up to me and very hesitantly said,

- *"but Jesper, all this doesn't make any sense".*

He looked at me with a bewildered expression. He was talking about quantum mechanics.

- *"I agree,"* I responded and continued, *"you are completely right. It doesn't make sense. it's good that you see it, just keep thinking".*

CHAPTER 5.

HORIZONTAL RAIN

"I've seen things you people wouldn't believe.
Attack ships on fire off the shoulder of Orion.
I watched C-beams glitter in the dark near
the Tannhäuser Gate. All those moments will
be lost in time, like tears in rain. Time to die."

— Roy Batty,
Ridley Scotts Bladerunner

Reykjavik, February 2004

It's dark and it's raining and I wish I would be anywhere but here.

There is no furniture. None. On the phone, the lady had assured me that the flat was furnished. *"Yes, of course, the rent includes furniture"*, she had told me.

And now I am standing here with my big duffel bag in an empty room lit by a single barren lightbulb and looking around. What on earth am I doing here?

I got here late last night. I took the bus from the airport and arrived at the house around midnight just as we had arranged on the phone and found it to be locked, with nobody answering at the door. The rain was horrendous, windy and icy cold, and I walked down to the center of town where a German chemist that I barely know lives and where I spent the night on a couch.

I spent the past two months in Copenhagen as a visitor at the Niels Bohr Institute with my employer, the Icelandic physicist Thordur Jonsson. My girlfriend and I split up in the fall; she left Iceland and went back home to Italy. The breakup had been underway for a while, it was my fault, our fault too, I guess. It was peaceful in a way, but painful too, like ripping off an arm. After she left, I stayed in Iceland a few weeks before flying to Denmark. And now I'm back on this island, where the sun never shines.

It feels as if I have come to the end of the world. Or returned to it.

I find a mattress in a built-in cupboard and roll out my sleeping bag. I'm going to work tomorrow. There is a meeting at the university and I am supposed to give a talk. I need to prepare for it. I get out my computer and sit down on the mattress with my back against the wall. I place the computer on my lap and sit there for a long while without opening it.

The lightbulb is broken, its light constantly flickering and it seems oddly synchronized with the rhythm of the rain cascading against the window. I look outside. There is a streetlight in front of the house and I can see the rain, it's horizontal, I've never seen rain like this. There is something incredibly eerie about this place, the constant darkness, all the concrete, and the rain. It gets to you. Today as I walked here, I saw some kids playing in the dark and rain. They played soccer.

The single flickering lightbulb cannot keep all this darkness out. It creeps in, the cold too, into this room and into my body, my bones and my unhappy soul. Without furniture, carpets, blankets, hot tea, and friends I cannot imagine how to stand it here. I have only spent a few hours in this room and it already feels unbearable. I put on my parka and my heavy boots and leave the light on, hoping that it will heat up the place while I'm out, and head for the center of town. I am going to pick up my favorite lamb sandwich at the food stand near city hall — it's medicine, I tell myself, I need it — and then I'll go to the Nordic culture house near the university. I sometimes hang out there; they have Danish newspapers and you can get a free refill on your coffee.

The institute in Copenhagen was very different from what I was used to. They all worked on string theory there and their only study group was on string theory too as were almost all the seminars I attended. I don't know much about string theory and I felt intimidated and very much out of place. Lost.

Before I left for Vienna four years ago, I spent a few months in Copenhagen, where I started working on a project on string theory for my master's degree, like everybody else. But I never really got it. It seemed complicated and technical to me, contrived too, I couldn't see its beauty and gave up after a few months. I decided to go to Vienna instead, where my girlfriend lived.

So I already knew a few of the professors in Copenhagen and had taken some of their classes. It felt different, though, to be at the institute as a visiting researcher rather than a student, people treated me differently I thought, they looked at me differently.

The Niels Bohr Institute is a maze. It was built in several stages, organically, without planning, house by house, and gradually connected by tunnels, corridors, and staircases till it eventually formed an incredibly beautiful web of connections and intentions and long-forgotten secrets. When I first visited the place I immediately fell in love with it. As a student, I sometimes went for a run in the nearby park, and afterward, I took a bath in the same bathtub that Erwin Schrödinger — the physicist who invented wave mechanics that later became quantum mechanics and who was given the Nobel prize with P.A.M. Dirac in 1933 — used when he visited the institute, or so the story goes. And later, when I became a postdoc there, I would sit in the famous auditorium A and listen with one ear to someone talk about string theory and extra dimensions and elephants on the moon and with the other ear listen to the walls that spoke of the passion, love, and aspirations of all the scientists who had sat there before me. I could see some of their faces in the pictures on the auditorium walls, faces of heroes, their names legend among physicists. The entire room is full of secret stories from the days when the new physics was still new, when quantum mechanics was first discovered, and the world was a newborn baby.

When I was finishing my thesis in the spring of 2002 in Vienna I had no idea what I would do next. I knew that I would not stay in academia — I simply did not have the confidence nor the scientific ideas for that. My education had been a confusing mix of five different universities in Denmark, Switzerland, and Austria, where I had pursued wild impulses, a wish to be near my girlfriend, to be in the Alps where I could ski and whitewater kayak, and a wish to be away from Denmark, which is flat and full of expectations. During my studies, I had learned a lot about theoretical physics, picked up a great deal of incomplete knowledge about different formalisms in quantum field theory and general relativity, but I hadn't built a basis to support a career and on top of that I had lost faith in my Ph.D. project.

The reason why I eventually did send an application to the University of Iceland, where I had seen an open post-doc position being advertised, was a combination of not knowing what else to do and knowing that Iceland is one of the most beautiful and interesting places I had ever visited. I even wrote in my application that I enjoy skiing and whitewater kayaking — something I was later told had been a decisive factor in getting the position since Thordur Jonsson and Laurus Thorlacius — the two professors in theoretical physics at the University of Iceland who hired me — thought that someone with an interest in outdoor sports would be more likely to cope with life in a place where you never see the sun during winter months and where strong winds regularly send cars flying off the road.

I had not found the move from being a Ph.D. student to working as a postdoc to be easy. I didn't know what problems to work on and felt lost. How do you create yourself as a scientist? How do you find your own direction, which way do you go? I couldn't just continue the line from my Ph.D. thesis as most postdocs do, in my mind that was a closed door. Instead, I tried working on a mix of wild ideas, where I kept returning to the questions I had been thinking about in Tibet, the questions concerning the foundation of my Ph.D. thesis.

So, I went skiing that first year. I did a lot of thinking, and let my relationship go down the drain.

✡

The wind is strong today. Sometimes I have to lean into it to make headway. I pass the city hall, a new building situated by the little lake near the center of Reykjavik and walk along the lake till I get to the tiny forest and the main road that connects the city center with the peninsula and the western part of Reykjavik. At the end of the road, there is a beach and a small white lighthouse that marks the beginning of the Atlantic Ocean. I walk past a little green spot with a few timid pine trees and long grass. The geese like this spot. I've seen groups of them block the busy four-lane road as they calmly march across the asphalt to find their favorite turf. I admire the confidence of these birds, it is as if the world belongs to them and they know it.

As I cross the road and get into the open I feel the full force of the wind. I love the smell. It's the ocean, you can smell it everywhere. It's alive. And the cold, I've always liked it cold, on ski tours, sleeping in the snow, I've always felt in my element.

Being outside lifts my spirit a little. It's not completely dark today, I can see the contour of the sun behind the clouds to the south. As I arrive at the cultural house I decide that I'm going to have a piece of cake with my coffee — that too is medicine.

CHAPTER 6.

THE RELATIVE

If I stand
alone in the snow
it is clear
that I am a clock

how else would eternity
find its way around

— Inger Christensen

According to the German philosopher Immanuel Kant, space and time are preconditions for our perception of reality. These are *a priori* concepts of the mind, which we bring with us into the world when we are born. Our experience of the world, our thoughts, and our perceptions are only possible because we are equipped with these concepts. According to Kant.

The fundamental role that space and time play for our way of perceiving and contemplating the world is the reason Einstein's theory of general relativity comes as such a blow to our sense of what makes sense. Kant may be right about space and time being innate concepts but in that case, Einstein's theory seems to suggest that those innate concepts are imprecise. It is as if the very fabric on which our thoughts are written is shaken by the work of Einstein.

And yet, anyone who spends time with Einstein's theory of general relativity and who has a fair knowledge of Riemannian geometry will be left with a sense of *"of course!"* There is something incredibly natural about this theory. Beauty is often mentioned in connection with theoretical physics and if there ever was a theory that deserves this label then it is Einstein's.

Put differently, if physics were music, general relativity would be Bach.

In fact, there are two theories of relativity. Einstein published the first one, the theory of special relativity, in 1905 and the second one, the theory of general relativity, in 1915.

The first theory is included in the second, where it plays the role of a local approximation. It is general relativity, that we need here. This is the theory that is generally believed to have a quantum counterpart. A quantum theory of general relativity. Or quantum gravity.

To say that Einstein's theory of general relativity is primarily concerned with space and time is actually slightly imprecise. What it is really about is the *geometry* of space and time.

So, the first piece of information we need to understand general relativity is that the world we live in is not flat. It has curvature, it has a geometry. This means that triangles in the real world do not usually have three angles that add up to 180 degrees. That would be true if they were triangles in a flat space — which is not usually the case.

The second piece of information we need is that the geometry of space and time is shaped by the energy and momentum *in* it. If we have an object with a large mass (mass equals energy) then it will cause space and time to curve. We say that the geometry is *dynamical* because it is not static, it changes.

What we experience as gravity is precisely this curvature. An object that is not accelerated will follow a path given by what is called a geodesic, which is what amounts to a straight line in this curved geometry. When Earth travels around the sun, it is not because there is some gravitational force acting at a distance between the sun and Earth but because the mass of the sun curves space and time, much like a heavy ball makes a dent in a soft mattress. In the same way, when we throw a ball into the air and watch it fall back towards Earth it is because the mass of our blue planet causes space and time to curve.

"What does it mean that time curves, too?" You might ask. *"You said that space <u>and</u> time curves?"*

Well, to answer this question let's first ask what it actually means when we say that the surface of something like a trampoline is curved? Imagine we have divided the surface of a trampoline into squares with a pen. Now, when we then stand on the trampoline and make it dent we will see that some of those squares become larger. This is because we stretch the elastic material of the trampoline. So, to make curvature we change the relative *distance* between points. To understand what curvature in space and time is, we need to think of time as just another dimension and when it curves the distance between its points changes in the same way. That is, the measurement of time, i.e., how *clocks* work and is why clocks run differently in different gravitational fields. When gravity is stronger – when there is a large curvature – time runs *slower,* relatively speaking, because we are stretching the squares.

This is something that can be measured. Although the difference is extremely small, two synchronized clocks that are placed at the foot and the top of the Eiffel Tower will not agree when they are compared later because the curvature of space and time is greater closer to earth. This is one of the amazing facts we have learned from Einstein's theory.

"But WHY do space and time curve?" I imagine you'd like to ask.

The answer is that we don't know. As is often the case in theoretical physics we can only say that Nature chose it to be this way. As if she had a choice, but we don't even know that, perhaps she had no choice?

In fact, I see one place we could look for an answer to the question of why relativity theory looks the way it does and that place is the final theory and the question of time. But we will get to that later, we need to cover some ground first.

✡

So what does the mathematics of general relativity look like? Well, first there is a four-dimensional space, which you can perhaps think of as the precondition that Kant wrote about. This space is not a space in the way we usually envision spaces, because it is not equipped with a *distance* between its points. It is what we call a *topological* space. For example, the topological spaces of a cup with a single handle and a donut are identical — they both have a single hole, and they are both finite. What makes a cup and a donut different is their geometry.

In general relativity, geometry is encoded in what we call a *metric field*. A metric field assigns to each point in the four-dimensional space a set of numbers, which tell us something about the curvature of space and time at that point. This gives us what we call Riemannian geometry. There is then a set of formulas known as Einstein's field equations that tell us how matter will affect this metric field, i.e., how *stuff* will cause space and time to curve.

So, this is the general setup: we have an abstract *topological* space and *fields* within it, and we have a set of formulas, that tell us how these fields behave dynamically, i.e., how they change in time. This is the setup that we encounter not only in general relativity but throughout modern theoretical physics. This is what is called a *field theory*.

✡

Now we have the foundation in place, quantum mechanics and Einstein's theory of relativity, the two grand theories that shaped much of the 20th century with their formulas: the nuclear bomb, the computer, and the discovery of the Big Bang. So far so good, now we just need the standard model of particle physics and beyond that lies the front.

If you listen carefully you can already hear the distant thunder of the canons.

CHAPTER 7.

THE FOUNTAIN

I'm a fountain of blood
in the shape of a girl.

— Bjork

Hekla, Iceland, May 2004

My phone is ringing.

I make a big sweeping turn just beneath a couple of huge blocks of dark lava protruding through the white snow and stop abruptly so that I end up facing south. In front of me is a magnificent view of the Atlantic Ocean. It's huge. It's dark with a million different shades of blue, and beneath it lies a barren desert of lava and sand painted in brown, red and yellow colors. It's a wild dance of incredible beauty — for a brief moment the view takes my breath away.

Niels, my Swiss friend, races past me on my right and breaks in an impressive cloud of white powder. You can tell that he is born on skis, no matter how much I practice I'll never find the grace that Niels has.

- *"What's up?"*

He looks at me and laughs. I can see the reflection of the sun in his goggles. I love the way he laughs.

My phone rings again. I pull off my glove and reach into my fleece jacket to get it. Incredibly, there is a connection out here. It must be because we're so close to the summit.

- *"Hi, it's Jesper,"* I say.

- *"It's Johannes. Eh, ... where are you?"*

- *"Hi, Johannes. Hmm ... I'm skiing. Where are you?"*

- *"I'm in Copenhagen. I thought we were supposed to meet today?"*

Damn! I suddenly realize that I have mistaken the dates and that I was supposed to meet Johannes in Denmark today. I thought it wasn't until tomorrow. I tell him that I have a flight later tonight and that I'll meet him tomorrow morning at the Niels Bohr Institute.

Niels and I have taken the day off and driven from Reykjavik to the volcano Hekla in my little Fiat Punto. We have spent most of the day crossing a huge lava field and have climbed the volcano on skis. It has been a gorgeous day. This is what I love about Iceland, this is why I came here. The air is incredible, the views, the snow, the solitude. As we're heading back down, there is not a single cloud in the sky. It's a perfect day and I should have been in Copenhagen to talk to Johannes about physics.

✡

I met Johannes in 2003 at a one-month workshop at the Mittag Leffler Institute just north of Stockholm. The topic for the workshop was mathematical physics and noncommutative geometry and I hadn't wanted to attend.

It was Thordur who had suggested to me that I should come with him to Stockholm for this workshop. I felt awkward, what was I going to do there, a whole month with a bunch of mathematicians? But I hadn't dared refuse him, so I sent off my application.

When we arrived at the Mittag Leffler institute I was given office space in the basement in the same room as Johannes, who had arrived a few days earlier. As the youngest participants at the meeting and the only Danish post-docs, it was probably inevitable that we'd start to hang out together. We both felt out of place, I think, unsure of our role and purpose in this place. Bored too, or at least I was.

So we began to venture off into Stockholm together to go to movies and drink coffee at cafes.

Oddly enough, we never discussed physics or mathematics during that month at the Mittag-Leffler. I remember it as a kind of unspoken agreement between us: we'd hang out, pass the time together and talk about anything but our professions. It felt like an embarrassing topic, which was best avoided. As if we both knew that we weren't meant to be there, that we were just pretending to be scientists and would soon be exposed and asked to leave — and by avoiding the topic we could save each other the embarrassment of having our bluff called out.

But then again, perhaps this was just the way I felt. Later, when I got to know Johannes, I realized that he was not as easily intimidated as I and that he was quite comfortable in his role as a mathematician. Perhaps the reason why he never asked me about my work or talked about his own was that he just hadn't thought that I would be interested in talking about it.

In any case, it wasn't until the end of the workshop when we sat in a cafe in Copenhagen and had a final coffee before heading off in our separate directions that I mentioned to him that I was interested in quantum gravity and that I thought that I had an idea.

It turned out that Johannes was very interested.

Johannes had completed his Ph.D. in mathematics a few months before our meeting. His thesis had been on the problem of defining what is known as a Dirac operator on a space with a boundary. As a mathematical problem, this is far from trivial but it is not the kind of problem that will revolutionize the world and Johannes was the kind of guy who knew this only too well.

The mathematician that I met back then was someone who appeared to have lost interest in his career even before it had begun. He seemed to be in academia only because he had not yet found the right moment to get off the bus. That he eventually would leave seemed inevitable, he just hadn't been bothered to sign himself out yet. It was not that he had lost interest in mathematics — I could tell that he cared deeply for his field and later, when we began working together and he ended up explaining all the advanced mathematics to me (over and over again), I also learned that he was a very gifted educator — but rather that he seemed to have little interest in the kind of problems that modern mathematics often deals with. That is, a very high level of abstraction and a limited focus on application to real problems. He wanted to work on something with an application, something that would make a difference also outside some narrow branch of mathematics. And he wanted to work on a problem where the answer wasn't known beforehand.

So the guy I met was open for anything that did not taste like the usual porridge. And when we started talking in the cafe in Copenhagen, he must have thought that the idea that I told him about was just the kind of crazy opening that he could be interested in. An opportunity that would make him stay in mathematics just a little longer and thus save him the trouble of figuring out what else to do.

So neither of us had any real prospects of a scientific career at that time, nor did we seek one. Looking back, I can't help thinking that this was the perfect condition for a new idea to take root. We didn't have anything to lose. We were not afraid of taking a risk because our careers were already dead, if not in reality then in our minds.

I had begun to search the literature in the summer of 2003 while in Iceland. I was looking for inspiration, something that matched the thoughts that I had while traveling in Tibet, something to guide me forward. What I did have was nothing but a loose idea. It was a hunch, or more than a hunch, but it lacked a concrete mathematical platform, a foundation, a language. So I read papers on different approaches to quantum gravity and hoped that I would eventually come across something that matched my idea.

And one day I found it.

I was sitting in my office reading papers on loop quantum gravity. I didn't know much about this direction of research, I had heard people talk about it, mostly negative stuff, *not to be taken seriously*, they had told me, but something made me read about it anyway.

And as I was reading a review paper I suddenly realized that this was what I had been thinking about in Tibet. Sort of, anyway. Close enough. And there was something else. They did it wrong, I was certain about that. In their analysis they did something that changed their approach from being an approach to a unified theory of quantum gravity — that is, potentially a theory of all four fundamental forces — to being an approach to quantum gravity alone. They cut something away, something essential. And they hadn't seen it. None of them had seen it, not yet.

I was so excited. This was a great idea! I could feel it in my stomach. This was it. Oh damn, it made me nervous too, when I get this out everybody is going to work on it and I'll be left behind.

I was wrong. Of course. Completely wrong.

✡

I told Johannes about this idea at the cafe in Copenhagen. And, as I said, he was interested.

He arrived in Iceland a couple of months later during the deep darkness of Icelandic winter. I picked him up at the Keflavik airport some 60 km outside of Reykjavik and drove him back to the city through the black lava fields. There is something incredibly ominous about this drive, especially during the winter. It's like a different planet, the hostility of the elements, the blackness, bleakness and wetness, the gale-force winds, the lonely glimmer from a distant house. On your first visit to Iceland you will be impressed, perhaps even intimidated.

I had sent Johannes some papers on loop quantum gravity and when he arrived we started talking about what to do about it.

It turned out that Johannes had actually carefully read the papers I had sent to him. Like, he had studied them. What I had done was something different. I mostly just skim papers. I avoid the details and just look for the main idea, that is my approach. Details are noise, they scare me too, but of course, if you want to get anywhere you need to understand all the details. And Johannes did.

So he began to explain to me what this was all about and how to understand loop quantum gravity. It was really exciting, I had found someone who was kind of the opposite of myself, someone who thought very carefully about the mathematical details.

What we wanted to do was to apply the mathematics of noncommutative geometry — a branch of modern mathematics pioneered by the French mathematician Alain Connes and which Johannes knew a lot about — to the general setup of loop quantum gravity. I had a rough sketch of how one might approach this and Johannes immediately picked up on the idea and laid out a strategy.

Sometimes I wonder if the idea I got in Tibet had in fact lost its way. It found me but it was meant for someone else. And that someone was Johannes. It should have been him. So I did the idea a favor, I brought it home.

But the truth is that this idea found us both. That is what I believe.

We published our first paper a year later and it was accepted in Communications in Mathematical Physics, one of the best journals in the field. I expected it to set in motion a tidal wave throughout the theoretical physics community — the naivety of youth I guess — but it barely caused a ripple. The sea remained as calm as ever.

And later, when I began to talk to people about our ideas, they mostly thought it was interesting, maybe even exciting, but clearly, nobody saw the revolution coming.

Perhaps ideas are always like that. They find someone willing to take them in, they lure you in with their vision, sweet words and promises, they catch you and stay with you. But they don't bother to lure in all the others, that's your job, your problem, once you've been caught it's up to you to pass it on.

It's up to you to free yourself.

✡

One evening, while I was in Iceland, I drove alone through the night along the southern shore just beneath the great Vatna Jökull glacier. I was heading east in my little Fiat Punto with studded snow tires. Throughout the drive the sky was lit with northern lights, green pulses dancing above me, luring me and my attention was divided between the snowy road ahead of me and the Milky Way above.

And then it all exploded. Suddenly the sky poured out over me and I stopped the car and got out, alone in the cold night and looked up.

It began in whitish-green, like a river above me, a spring, a fountain. I stood *above* the fountain and looked down into it as it poured out over me, over the entire world around me, where I was the sole inhabitant. I stood at the bottom of the road with my back against my car and my neck against its roof and everything poured up towards me, passed me, in all directions, sweeping across the snowy hills like a blizzard of beauty and with me in its center.

And then the fountain exploded again, with new rivers, new springs that all reached up towards me and the colors changed, whitish-green, then green, but with springs of red, yellow, blue; faster pulses, like a symphony in light, with themes constantly flowing into new themes, intertwined and yet so simple. It was Bach and it was Bruckner and it was completely silent.

And then it was over. The spring above me dried out and I was left alone in the night. I felt that it had filled me, that it was my spring and that it had sprung for me.

And from me.

CHAPTER 8.

THE STANDARD MODEL

"Maybe nature is fundamentally ugly,
chaotic and complicated.
But if it's like that, then I want out."

– Steven Weinberg

In fact, we already have a theory of every***thing***. It is called *the Standard Model of Particle Physics* and is a theory of all the *things* that we know of.

With 'things' I mean all the elementary particles that we observe in Nature and all the forces by which they interact — except gravity.

When you look into an atom you see two things. There is a nucleus and there is a number of electrons, which sort of fly around the nucleus. The reason I write 'sort of' is that we are now in the domain of quantum mechanics, which means that things are not the way we are used to — as we have already discussed — but for now it's enough to know that there is a nucleus and electrons. If we look into the nucleus we find a number of protons — depending on which element we have — and a number of neutrons. This is the domain of nuclear physics. If we then look into the protons and the neutrons we find quarks and gluons. This is finally the domain of particle physics.

To understand what the standard model of particle physics is all about we need to understand a mathematical framework known as *gauge theory*. And to understand what a gauge theory is we start with a space — say your living room. Now, imagine that you want to move something from point A to point B in this space. If you simply want to move a round ball then this is straightforward, there are no mathematical ambiguities involved, you just move it. But if the object you wish to move is a stick, that is, something with a shape, then it is not completely clear how to proceed. At point A the stick points in some direction, so when you move the stick to point B in what direction should it point there? The same direction or somewhere else?

The point is that to determine this you need some extra information. This extra information is called a gauge field. A gauge field tells you how you should orient a stick if you move it between two points in a space. It is like a cooking recipe for moving sticks between points in your living room.

Now — and here it gets a little tricky but stay with me — you can also have a stick that lives in some internal, abstract space that has nothing to do with the space in which you move — say, your finger — between points. Imagine having an arrow attached to a piece of paper with a needle. When you move your finger from one point in space to another you can simultaneously turn the arrow around the needle on the paper. In this case, the internal space has two dimensions — the dimensions of the paper — while you stand in a three-dimensional space (your living room). All this adds a level of abstraction to the setup, but the math is the same, you still need a gauge field to tell you how to move the arrow.

This is about symmetry and it is about what we call a *local symmetry*. A gauge theory is a theory that has these internal sticks that can be turned in different internal directions at each point in space (and time). The crucial point is that the physics of the theory does *not* depend on any particular direction in these internal spaces — and this is a *local* symmetry in the sense that we can turn the sticks at each point in each and every direction and the physics stays the same.

It turns out that Nature likes this setup. For instance, electromagnetism, which is the theory that describes electricity, light, radio, x-rays and the entire spectrum of electromagnetic radiation, is a gauge theory. This means that the electric and magnetic fields can be combined in a certain way to form a gauge field that corresponds to a two-dimensional internal space.

In fact, all the three forces of Nature that are described in the standard model — the electromagnetic force, the strong nuclear force and the weak nuclear force — are all formulated as gauge theories. These gauge theories correspond to internal spaces that are two, three and four-dimensional.

The standard model of particle physics is the mathematical theory that tells us how these three gauge theories interact with matter. The theory involves a number of different particles and a special mechanism that generates mass to these particles — the famous Higgs mechanism. The whole thing comes in three duplicates, called generations, which are identical up to particle masses.

But why did Nature choose this framework of gauge fields? Why are there three local symmetries, where abstract 'sticks' can be turned in arbitrary directions? I imagine that you'd like to ask me now.

These are good questions. And the answer is that we don't really know. We would like to know — this is one of the key questions being asked in modern theoretical physics. We have ideas concerning the origin of the structure in the standard model and one of these ideas — which is intimately related to the mathematical research field known as *noncommutative geometry* — will play a key role in the following chapters.

But before we get to that we need to complete the story about the standard model because what we have discussed so far is not the full picture. The point is that the standard model is also a quantum theory. This means that we need to replace physical quantities that represent *numbers* with their corresponding *operators*. This is the process of quantization as we discussed in chapter 4. In the case of the standard model we are not dealing with ordinary quantum mechanics, where you replace quantities such as position and momentum of a single particle with corresponding operators. Instead, we now have *fields* (for instance, the gauge field). The standard model is a field theory and that makes its quantization much more complicated. The quantization of a field theory — called *quantum field theory* — is a highly non-trivial undertaking.

Let us pause here for a moment to consider this (and if you wish to skip the next few pages be my guest, this is not essential for the rest of the book). The study of quantum field theory is a highly advanced field of research, which is still being developed. Thousands of researchers have spent their careers on this topic and have developed several different mathematical formulations and approaches. And yet, despite this tremendous effort we still do not know how to construct a quantum field theory for a physically realistic model such as the standard model. That is, a quantum theory in the sense of what we discussed in chapter 4 with an algebra and a Hilbert space.

We can make such a quantum field theory for special cases, certain toy models, but *not* for the standard model.

What we do have is something slightly different. We have an *approximative* theory. That means that we have a mathematical framework of approximations that is capable of producing numbers, which can be compared to outcomes of actual experiments. This is the kind of quantum field theory that the standard model is. We can compute the numbers we need for a comparison with experiments of ever-higher precision just by adding more and more terms to our approximation. So, if we want a number that is accurate to, say, the 20th decimal place, we just need to crank our computational machinery to get it.

But what is this approximative theory an approximation of?

Well, that's the question. At a heuristic level, it is an approximation of a quantum theory — an algebra and a Hilbert space — but that is only true at a heuristic level. The point is that we know that the approximation — which is an infinite series of mathematical terms of increasing complexity (and which we in principle know how to compute, although in practice it gets very hard very soon) — will give us better and better numbers up to a certain point and from there on the approximation will begin to diverge. That is, our approximation does *not* converge. If we could add up all the terms it would be infinite.

"But that sounds crazy!" you might add. Well, yes, that is true, but the point is that all of this is of no practical consequence. The point where our approximation begins to diverge is nowhere near experimental data and never will be. Nevertheless, the fact remains that we do not have a theory outside of this approximation.

In technical terms, we say that we do not have a *non-perturbative* theory.

Now, there are different approaches to deal with this awkward situation. The obvious one is simply to promote this approximation to the theory itself and simply understand that quantum field theory is a perturbative theory. Another option is to say that our theory is a limit of another yet unknown, theory that operates at even higher energies.

So, to recapitulate, what we are left with is a highly effective and for all practical purposes sufficient mathematical machinery that can stand the test of experimental verification. And that is pretty damn good — we should be absolutely clear about just how impressive a theory the standard model really is. Its mathematical structure has been confirmed by numerous spectacular predictions, the latest being the confirmation of the existence of the Higgs particle and the quantitative predictions of this theory have been tested to incredible accuracy. The standard model has never failed.

Nevertheless, the standard model clearly cannot be the final answer. In addition to the issues mentioned, it does not include gravity. Also, its particular mathematical structure raises a long list of questions. Why, as already asked, this choice of internal symmetries, the gauge structure? Why three generations? In fact, the entire mathematical structure appears to be one big question mark. Why does it look like this? It does not appear to be an arbitrary structure; it has a very particular form. It looks like a question that *has* an answer.

On top of all this, it involves a number of parameters like coupling constants and masses that come without any explanation except that those are the numbers that fit the experimental data. This is not exactly what we expect a final theory to look like.

As I have already mentioned, it is widely believed that the standard model is a low energy approximation of something else. What this something else might be is not known. That is the point.

So this is what it all boils down to. We have a theory of gravity called general relativity and we have a theory of particles, called the standard model. These two theories represent — with a few particular exceptions — everything we know about the world, understood in a reductionist sense. These theories are our ultimate experimental data.

The exceptions I just mentioned? Dark matter and possibly dark energy. These are both important but I will not spend time on them in this discussion.

✡

We are now getting close to the frontier. We are zooming in on the point, where what we know is behind us and what we don't know is ahead of us. But before we step out onto the battlefield, before we leave the outermost trenches and charge into combat, we need to take a deep breath, get a good grip on our rifle, mount the bayonet and count to ten.

CHAPTER 9.

A POETIC UNIVERSE

When I heard the learn'd astronomer,
When the proofs, the figures, were ranged in
columns before me,
When I was shown the charts and diagrams, to add,
divide,
*   and measure them,*
When I sitting heard the astronomer where he
lectured with
*   much applause in the lecture-room,*
How soon unaccountable I became tired and sick,
Till rising and gliding out I wander'd off by myself,
In the mystical moist night-air, and from time to
time,
Look'd up in perfect silence at the stars.

— **Walt Whitman**

One of the most remarkable things, that we learn from physics is the fact that everything that we observe in the world around us has already happened. All of it, the stars that we see in the sky, the sun that feeds us, the trees outside my window that slowly sway in the evening breeze, the computer that I use to write these words. It is all written in the past tense.

The only 'now' that exists in this world, that we observe, is in our heads. All the rest lies firmly in the past.

And the further away we observe an object the farther back in time it lies. The coffee cup that stands on my table happened close to a millionth of a second ago — it is almost now, but almost is not now, it is solid past. The moon that I see rising on the evening sky has already happened, something like a second ago, and the sun is even further in the past, a whole 8 minutes and 20 seconds.

How beautiful this is, this reality of ours that has wrapped itself in the past tense.

The point is, of course, that light has a speed, a finite speed, which means that it takes time to see. It takes time for light to travel through space and reach our eyes with information about the world and everything in it.

And here is what makes my heart beat faster. This Universe had a beginning — we all know that, or we think we know it, which is all we can do, to think what we know — a first moment, which means that the further we gaze out into space the closer we get to this first moment. No matter in which direction we look we eventually meet this barrier, the beginning of our Universe. It is like a huge enclosure in space and time, the Universe that we observe has wrapped itself in its own beginning, its own birth.

I find this immensely poetic.

The farthest that we can see when we gaze deep into space is the moment of recombination. Its name is a little misleading, it refers to the moment when atoms were first created some 380 thousand years after the Big Bang. It should be 'combination', the "re" is superfluous. We think.

At the moment of recombination, the Universe became transparent. Before the formation of atoms, it was opaque, it was like trying to use a flashlight in a bowl of milk, but once atoms were created light could travel freely. And the light that was set free at recombination is still with us, it is everywhere, like a faint shadow of the Big Bang. It is the famous cosmic microwave background radiation, which is now a major field of research.

But think about this. At every point in the Universe today, there exists the faintest image, the faintest memory of the Universe's own beginning, its *first childhood memories*.

I often think of the microwave background radiation as a beautiful metaphor for our own lives. We grow up, we become adults, we live adult lives. But we never fully escape our beginning. Our first childhood experiences always linger somewhere in the shadows of our consciousness, as the background of our lives.

We cannot look past the point of recombination with ordinary telescopes. No electromagnetic radiation was able to travel freely from a point in the past that lies before the moment of recombination. It is conceivable that one day we will be able to see past this point by other means, perhaps via waves in the gravitational field. But as of today, that is science fiction.

If, however, we could look past this point we would soon enter the domain of quantum gravity. How so? When we look out into space we look into the past, as I said. And in the past lies the Big Bang where the Universe is believed to have been contracted to a single point of immense curvature and density. This means that we reach domains of higher and higher energies, regimes where our present laws of physics no longer apply, the domain of a quantum theory of gravity.

This is remarkable. Science has always walked in two separate directions: the search into the very small and the search into the very large. The reductionist approach that goes from chemistry to atomic physics and all the way down, and the approach of astronomy and cosmology with observations of the stars and deep space. Now these two meet up at last. Both of these directions of research bring us to the exact same point, the same question, namely that of quantum gravity and the possibility of a final theory.

I say it again: what a poetic Universe we have.

CHAPTER 10.

AT VERDUN 1916

"Hard pressed on my right.
My center is yielding.
Impossible to maneuver.
Situation excellent. I attack"

— Field Marshal Ferdinand Foch

Copenhagen, Denmark, June 2006

If they continue like this, I will soon be unable to breathe. It's like a battlefield. We are at Verdun in 1916 and the Germans are attacking.

I start over again. I wipe the blackboard clean with the eraser and pick up a new piece of chalk.

Both Ryszard and Adam are smoking. A lot. They sit in each of their corners, Ryszard behind his desk and Adam in an armchair by the window. We have been here for more than an hour and a half and they have been smoking the entire time.

Adam is mostly silent. He watches me with his eyes squeezed to a thin slit. Ryszard sometimes interrupts me with a question or with an impatient *"yes, yes, of course, go on!"* when I explain something that he has already understood.

We are in Ryszard's office in the department of mathematics at the University of Copenhagen. He is a professor here and Adam is a postdoc in his group. I am standing by the blackboard trying to explain to them what Johannes and I are up to.

Ryszard was Johannes' Ph.D. advisor. A few weeks ago, he heard Johannes and I talking about our project during a coffee break at a seminar and asked me to come by his office and tell him about our ideas, what we're working on.

I got a job at the NORDITA institute, a three-year postdoc position at a section of the Niels Bohr Institute that is under the Nordic Council of Ministers, and now I'm here, in Copenhagen, back in Denmark after six years abroad.

I don't yet know enough about Ryszard to be intimidated. During my brief period as a student at the Niels Bohr Institute, I did sit in some of his classes, but I never realized who he was.

I never realized just how intelligent he is. Not until now.

I draw a new graph, connect it to the chain of graphs that I have already drawn on the blackboard and explain once more our point about gauge connections. I quickly glance at Ryszard. What is his reaction? What is he thinking? This is the point where we involve noncommutative geometry, this is where I enter his territory. He is the world-renowned expert. I have the sensation of walking on the edge of a razor and I can't help thinking that

Ryszard understands what I am talking about better than I do.

This is a sensation that will come back to me often in the years to come. When I tour the world with our ideas and try to convince other scientists of our work, Johannes and mine, I seek out both physicists and mathematicians. And whenever I stand in front of an audience of mathematicians I get this feeling again: that one or two of the mathematicians in the audience get it, that they understand the math, the construction, and that they understand it better than I ever will.

Ryszard's grey hair stands right up from his head and his gestures are hectic. He is interested, I can tell. I like his energy, his passion. He is intense. Very. And I am impressed with how quick he is. He immediately picks up on what it is we are trying to do and he starts to fire ideas back at me.

Good ideas, too.

Afterward, when we sit in the cafeteria on the top floor and drink a cup of coffee, Adam tells me that it is *career suicide* for a postdoc to work on a project like ours. Way too hard, way too risky. "*It's crazy, mate*" he laughs and lights another cigarette.

Adam is Australian, he has a big beard, a ponytail, and a smile in his eye. I like him.

✡

Oberwolfach research center, Germany, December 2005

This room is perfect. It has a huge window facing the forest and in front of the window stands a desk with a simple chair and an elegant lamp. The room has a single bed and a closet. Nothing else. No computer, no phone, no details.

The silence in this room is more than the absence of sound. The design itself is silent. The straight lines; the simplicity; its energy, calm and serene. Nothing speaks here. There will be silence in this room no matter how noisy it may get.

I've just arrived at the institute, I took the early train from Copenhagen and arrived at the Wolfach station in the late afternoon, where I was picked up by the institute staff in a minivan.

On the final stretch of the journey, I had begun to notice a couple of people around me. Little things, the way a jacket was worn, a thick shoulder bag that could no longer be properly closed, a pair of heavy glasses, eyes, a particular look. These were theoretical physicists, I could tell, and these were also the people who got off at Wolfach and stood patiently and waited until the minivan arrived. They were clearly professionals, conference veterans who knew the drill.

I put my bag on the floor of the room and get out my computer to check my email, but I soon realize that there is no Wi-Fi here. I check my phone: no connection. This place is really serious, they mean business. I head down to the dining area where I noticed some tables with newspapers when I first arrived — since my time in

Vienna I have the habit of reading the main German newspapers whenever I can.

On the way down the stairs, I meet Mario carrying a suitcase. I met Mario for the first time a few months ago at a conference in Potsdam, Germany, where I gave my first talk on our work. Mario is a German theoretical physicist. He is not very tall, has round glasses and a friendly face. In Potsdam, he had been very interested in my talk, had read our paper too, and he had told me that he was one of the organizers of an Oberwolfach workshop a few months later and that he would like me to give a talk there as well.

He now greets me with a smile

"Hi, Jesper, great that you made it. It's good to see you, how was your trip?"

He tells me that I'll probably be asked to talk on Thursday but that he still needs to arrange a few things with the other organizers so that might be changed. He promises to give me plenty of time to prepare.

After my chat with Mario, I decide to check out the lecture halls and the library. This place is famous, and I want to explore a little. I walk out of the residence building and cross the little garden with the *Boy Surface* sculpture — I recognize it from pictures. Someone has told me that there used to be a small castle here in the middle of the Bavarian Black Forest but that it was demolished in the 1970s to make room for the new institute of mathematics that was built in its place. It may have been a shame about the castle, but the building that I am now entering is truly beautiful. It is perfect for mathematics. Its design is very pure, like the horizon at the Danish west coast, it is perfect. There is a serene atmosphere, it is a temple of deep contemplation.

The building's interior has the same calmness as my room. The walls are covered with dark wooden panels, the floors with carpets that dampen the sound, and the light comes from hidden sources in the ceilings.

Across from the entrance hall is a series of large panorama windows with views towards the valley. I think I recognize the two chairs placed in front of them, they're by some famous designer, perhaps Danish. I turn right and walk into the large lecture hall, where I pause for a moment by the door to take in the atmosphere. I'm the only one in the building; no sounds anywhere. There are 50 chairs lined up in five rows, all with black leather seats and all facing the three large double blackboards at the end of the room. There are two large floor-to-ceiling windows with views of the forest framed by white drapes. Here too, the walls are covered by dark wooden panels. I notice a little washbasin to the right of the blackboards, it's an elegant design, a black beam of steal wrapped around the basin.

I go downstairs where the library is, — there is also a small room with a telephone, I notice — and later I also find a music room with a large piano. It feels like walking in an empty church.

I've heard a lot about this place, the Oberwolfach mathematical research center, and all its traditions. That there are coffee breaks with cake every day at 3 o'clock in the afternoon; that there is a traditional Wednesday hike to a nearby village, where you always order the famous *Schwarzwälder Kirschtorte*. Someone told me about their way of serving dinner, where you are assigned a new seat every evening at one of the twelve round tables to ensure that you get to meet all the participants of the workshop. And I've heard that workshops here never have a fixed program, but that people are instead asked to contribute according to how the workshop proceeds.

And here I am, at one of the holy temples of modern mathematics. My feelings range from awe and excitement to intimidation and anxiety. But of all the places in the world, this is the place I most want to be.

During the Thursday morning coffee break I walk up to one of the organizers and ask him how much time I have for my talk. *"60 minutes"*, he replies — I knew that already, all the talks are 60 minutes — but I tell him that I actually only need 45 minutes for my presentation, *"would that be okay?"* I ask him. *"Sure"*, he says and when he introduces me a few minutes later, he tells the audience that the next coffee break will be 15 minutes longer.

I've practiced my talk over and over again, 45 minutes exactly. First at my little blackboard in my office at the Niels Bohr Institute in Copenhagen and later in my mind during walks in the forest around this institute. I know it by heart — but I don't know which questions people will ask me.

I start talking about our ideas, about how we aim at combining the mathematics of noncommutative geometry with quantum gravity, that we want to use elements of loop quantum gravity in a new way. I draw a lot of graphs on the blackboard. I don't just run through my slides but bravely venture out onto the floor armed only with a piece of chalk and a pointer.

I can feel the audience. They are friendly, they are mostly European mathematicians and theoretical physicists, many are German. During the week I've listened to most of their presentations, so I know what projects they work on. There have been many talks related to the subject of my Ph.D. thesis, this old idea of Pauli's, and that makes me feel oddly confident, as if I possess a secret knowledge about their work, that it's wrong and that only I know it. I also know that most of them are probably better physicists than I am, their technical skills are beyond mine. It's a strange feeling, as if I *see* it but don't *know* it.

After my talk, someone from the audience approaches me.

"I liked your talk a lot", he tells me and continues, *"it's interesting stuff. You know, I was thinking, would you like to sit down afterward for a chat? I'm interested to hear more about these things that you and Johannes are working on."*

Of course, I accept, I'm flattered that he wants to talk to me. As he walks off, he remarks with a smile:

"you know, your talk was the most interesting in the whole conference, well done."

At the 3 o'clock coffee break I meet up with the man and we bring two large pieces of cake and two cups of coffee with us outside and sit down in the grass. It's a sunny day and quite warm. Many of the conference participants are scattered around in the small garden. The man, a Dutch professor, has brought some paper with him, but before we begin to talk about physics, he asks me a few questions about how my collaboration with Johannes came about and where we're both working.

He's one of the younger professors, but apart from that I don't know much about him. He is very friendly. I take the paper he hands to me and begin to write. I first draw a couple of graphs and explain to him that the graphs we use should only be thought of as an approximation, as a tool that we use to search for an underlying theory. We don't yet know what that theory is, but we think that we can use the graphs to figure it out. He listens attentively, it's much easier like this, just one-on-one, no audience that makes my brain freeze, just a single guy who means well.

He asks me why we don't just use the loop quantum gravity setup and I explain:

"I think that people in the loop quantum gravity camp have overlooked this. If you change their construction in this way..."

I write a few formulas and show him

"... then you're automatically moving towards a construction that resembles what Alain Connes is doing. That is what we want, we want to turn their theory of quantum gravity into a theory that also involves the standard model of particle physics — or at least something that is more than just gravity."

I like the way the Dutch professor listens. He clearly doesn't care *who* I am, he cares about *what* I have to say, he's interested. That gives me courage, so I begin to talk about the philosophy behind it all, what I think makes this interesting.

"I think a final theory must be founded on some sort of principle, that is essentially empty. Like, when we go deeper and deeper into all this ..."

I wave my hand around to indicate what I mean

"... and find increasingly abstract theories, that are based on simpler and simpler principles, well, where will that end? I think that this process of reduction must end with something that is almost empty, like, there can be nothing there. It's going to be almost trivial. What else could it be? It has to be something that cannot be reduced anymore."

The Dutch professor looks at me. This is the reason why we want to look at loop quantum gravity and the holonomies, which is the basic mathematical building block they use. I continue

"When the loop guys take the trace of the holonomies..."

I write the formula down, show him where in loop quantum gravity they change everything by doing what mathematicians call 'taking the trace'. I continue

"... they throw away something essential; if they didn't do that they would automatically land in the domain of Alain Connes and his noncommutative geometry. So, quantum gravity already has a mechanism of unification within it. This is what I find so fascinating. It's like all the gauge degrees of freedom, all the forces that we find in the standard model, they could be just a

spin-off of pure gravity. I think this would be incredibly beautiful."

- "Hmm ... I see, yes, that would be nice. But what exactly is that simple principle that you get if you don't do as the loop guys do?", he asks me.

- "I don't know yet, we're not sure, but it's like when you move stuff around in space, — I mean, the holonomies — then there is this extra structure ..."

I move my hand through the air and twist it in an attempt to show him what I mean

"... as I said, that looks a lot like the structure that Chamseddine and Connes work with. But we're not sure exactly how to formulate it. Not yet, it's just an idea right now. But we think that the lattices can help us find it."

We sit in the sun for a while talking. Although it's November it's not cold. When the bell rings to announce the last two talks of the day we get up and join the others.

South coast of Iceland, May 2004

Jon looks agitated. I watch him from the car as he walks around the parking lot speaking on his phone. I can hear some words, but my Icelandic is not good enough to understand what he is talking about. I can tell from the look on his face that it's good news.

He jumps into the driver's seat and starts the car.

\-   *"There are huge waves just off Hveragerdi. The others are already there."*

He drives out of the parking lot and continues

\-   *"it's the storm that passed a few days ago, the waves are coming in now and they are massive!"*

He doesn't even pause to ask us if we want to come with him. Of course, we do. We're off to Hveragerdi to kayak the biggest waves of our lives, seven meters at least.

I watch in awe as the wave engulfs one of the youngest guys — I pulled out at the last moment. It's too big, the wave. I watched it from my kayak as it came in, a dark and ominous shadow growing just beneath the horizon and I could tell that it was a monster. Now it breaks on top of Eigil, who is paddling like mad. It's a mountain of water and he is crazy enough to surf it. He disappears completely for a second and then he is shot out through the wall of cascading water, like a rocket, a short kayak with a little man shot out in front of the line of a breaking wave. He is literally airborne. It's the most incredible sight I have ever seen, the kayak bounces wildly up and down in front of the wave as it races towards the shore.

Surfing seven-meter waves is not something I have tried before. In theory, it's very safe but safe doesn't quite feel safe when a giant wave towers up behind you ready to break on top of you.

The trick is to place yourself in just the right place — too far out and you'll miss it, too far in and it will crush you — and just when the wave begins to build behind you, you paddle hard to pick up the speed necessary to ride it. If you don't have enough speed the wave will flip you over, it will cartwheel you forward and break onto you as you are upside down and push you down towards the bottom. But if you've got enough speed, you'll catch the wave, or the wave will catch you, and you'll ride it to hell — or so it seems.

I sit in my kayak a little further out and watch the horizon. The waves come in sets and right now there is a lull. I can see the others in their kayaks, yellow helmets bobbing up and down in the dark choppy waters, all watching the Atlantic with anticipation, moving their kayaks slowly back and forth trying to find the right spot.

Suddenly, I see a faint shadow building just beneath the horizon. It's hardly anything but it's there. And then there is another shadow, it's a new set of waves coming in. I decide to let the first one pass and go for the second, it looks more regular. As the first wave grows, I paddle outward to let it pass and place myself for the second wave. It's big. As it moves toward the coast it looks like a sea monster swimming just beneath the surface, a creature from the depth of the ocean moving toward us full of rage.

When the wave is perhaps 40 meters behind me, I begin to paddle. I don't look back, but I can sense it, a massive wall building behind me. I paddle with everything I've got as it builds and builds. My kayak picks up a lot of speed as the water steepens and then I no longer need to paddle, my kayak darts forward.

Everything happens incredibly quickly, my kayak begins to bobble from side to side in the steep water and threatens to suddenly veer off to one side and throw me into a sidewards roll in front of the wave. But I manage to keep it straight, riding the wave, racing down the ocean.

And then the wave breaks, the massive tower of water crashes down on top of my kayak and engulfs me. The power of the wave catapults me forward. It's a rocket boost that is almost impossible to handle and after a few frantic seconds I'm really out of control. My kayak is cartwheeling crazily in the breaking wave in meters of white water, where up is down and then up again and I twist my paddle and try to roll up, but there is no up, just white bubbles and black water and gravity in all directions.

And then it's over. The wave pushes me onto the beach where my kayak bounces on the rocks and my hands begin to bleed and it feels great.

Let's do it again.

CHAPTER 11.

WHY DO WE NEED A NEW FUNDAMENTAL THEORY OF NATURE?

"We must know.
We will know!"

**— Professor David Hilbert
Königsberg, 1930**

It's time to get out of the trenches.

From the material, that we have covered so far — quantum mechanics, general relativity, and the standard model — two central questions emerge.

First question: *Does a quantum theory of general relativity exist?*

Of the four fundamental forces of Nature (the electromagnetic, the weak and strong nuclear forces, and gravity) gravity stands out as the only one that we do not know how to quantize.

The quantization procedure that works for the standard model and a large number of other field theories does not work for general relativity. When we try to apply it, we run into problems with what is called *renormalization*. Renormalization is an important procedure in quantum field theory that removes certain divergencies, which emerge in the quantization procedure and it does not work for gravity. It's not a settled issue that a quantization of general relativity within the framework of quantum field theory is impossible or meaningless, but as of today, that does seem to be the case.

"But do we know that gravity should be quantized?" you might ask. The answer to this question is *"No, we don't."*

This is a critical point that I would like you to remember because it will be important later on in this book: we have *no* direct empirical evidence that gravity must be quantized.

What we do know is that gravity interacts with matter, which *is* quantized. Einstein's field equations have on the one side the curvature of space and time, and on the other side the energy content of matter. Classical quantities equated with quantized quantities. This means that *something* has to happen. You cannot have quantities that are described within a quantum theory, and quantities that are not, appear in the same formula.

We also know that exotic objects and phenomena like black holes and the Big Bang are a part of reality and that they involve curvatures of space and time so great that we end up in an energy regime beyond the Planck energy. This is a regime where some sort of a quantum theory of gravity is expected to apply.

However, it is important to realize that gravity is not a force like the other three. Its mathematical structure and the role it plays are of a completely different character.

Gravity provides the *background* on which everything else happens. When we quantize the standard model, we do so with reference to this background, that is, with reference to a given geometry. General relativity is thus completely intertwined in the standard model as the provider of the geometry in which the other fields live.

What this means is that the standard model is *background dependent*, it depends on a background provided by general relativity.

This is the beginning of a major headache, because if we are to quantize general relativity, — i.e., the background itself — with respect to what background should we do it?

Clearly, there cannot be such a background. If we are to quantize geometry itself, there cannot be a *second* geometry to provide us with a background. Thus, we encounter the issue of *background independence,* which simply means that we need a quantization procedure that is independent of a background geometry. This is a crucial question that has been wildly debated for decades. We will return to this issue later.

Keeping in mind the special character that the gravitational force has, it should perhaps not come as a surprise that the standard method of quantization fails in its case. Indeed, one could turn this issue around and say that *if* we had been able to quantize gravity in precisely the same way as the other forces, then we would have been left with one *less* clue as to what a final theory might look like.

This leads me to a general consideration. Einstein's field equations bring together two quantities, which are a priori of a very different nature. The curvature of space and time on the one hand, and the energy content of matter and forces on the other.

Perhaps these equations should be read as an indication that these two quantities are not as different as we think they are. Perhaps the geometry of space and time and its content should, ultimately, be seen as different manifestations of one and the same thing? If this is the case it should entail an entirely different interpretation of what quantum field theory really is.

This is the kind of reasoning that one also comes across in the research field known as *noncommutative geometry* in modern mathematics. We will learn more about this in a little while. But first, we have the second question.

Second question: *Why does the standard model of particle physics look the way it does?*

This is what we would truly like to understand: so, we have a theory of everything (as I said, the standard model, a theory of everything except gravity) and it took an immense amount of work to find it, the number of people who dedicated their lives to its discovery is huge — in a way it took several thousand years of scientific development to uncover it — but now we have it and it works wonderfully, truly, truly wonderfully — but it looks so damn *weird*.

Why does the standard model look so weird?

Well, in a way we know why: if it had looked any different, we most likely wouldn't be here to ask these questions, — *that's why!* This is the anthropic 'why', the world is the way it is because we wouldn't be here if it was any different.

But that is not the why we want. We want a real explanation, one that actually tells us *why*. We want to *know,* and we will not make do with semantics.

So, let's ask again: why does the standard model have its very particular structure, where three different gauge groups that correspond to internal rotations in 2, 3, and 4 dimensions, are stitched together in a very specific manner that involves a Higgs field and around two dozen free parameters, such as coupling constants and masses, everything neatly wrapped in quantum field theory? And why does the whole package come in three almost identical copies, the three particle generations?

The problem with the standard model is that on the one hand, it looks very contrived, there is something almost arbitrary about it, but on the other hand, its mathematical structure reveals subtleties that suggest that it is anything but arbitrary.

In theoretical high energy physics today, there is a tremendous amount of fiddling around with variations of the standard model and gauge theories in general. And of course, this experimentation is essential, we need to peek into all possible alleys for clues as to what lies beyond the next hill. But this fiddling around cannot be all we do, we also need to ask the bigger questions, which is *"why gauge theory"* and *"why quantum field theory"* — and why this strange combination called the standard model. There is of course string theory and that is *one* idea, but we need more ideas. We need all the ideas we can get.

It is often said that we do not have experimental data to guide us in the search for a unified theory of quantum gravity and that is strictly speaking correct — we have no experimental data that points directly towards quantum effects of the gravitational field — but I believe that with this point of view we overlook the mountain right in front of us: that the standard model *itself* is a huge piece of experimental data (together with general relativity) that we need to first interpret and then explain — and this explanation can only come in the form of another, deeper theory.

This implies that the theory we are searching for has a daunting task to complete: it must be possible to *derive* the standard model — or at least its overall structure — from it.

So, the answer to the question in the title of this chapter *"why do we need a unified theory of quantum gravity"* is that we need such a theory to provide us with answers to these two questions: the question of quantum gravity and the question of the standard model, which in their totality include, literally, *all of it*.

It may very well be, however, that the answer to these questions does *not* involve quantum gravity and does *not* involve a unification of the fundamental forces. Who knows? It could be something completely different. This is the nature of research. It is the DNA of adventure. When you enter that forest you do not know if you will find what you search and, in fact, you don't even know exactly what you are searching for.

But whatever it is, we want to know it.

This is the most fascinating intellectual problem you can possibly imagine. What a challenge! And what a gift to be given the opportunity to chunk away at this problem, knowing that of all the gold diggers out there it might as well be *you* who finds the diamond.

CHAPTER 12.

KAFKA'S LOCK

"Klamm will never talk to you.
How can you even think of such a thing!"
"And won't he talk to you?" asked K.
"Not to me either," said Frieda,
"neither to you nor to me,
it's simply impossible."

— Franz Kafka, The Castle

Kafka. Let's pay Kafka a visit in his Castle — or rather, let's visit the nearby village where one of his main characters resides. He's got something there that I would like you to see.

In his novel "The Castle", Kafka tells the story of the land surveyor K. who has been summoned to a castle for an assignment. However, soon after he arrives in the village at the foot of the castle hill, K. learns, by way of a letter from the castle officials, that he has been erroneously requested. He isn't needed there and there is no assignment. His summoning was a mistake.

But the land surveyor in Kafka's novel is not satisfied with this lack of clarity and explanation and thus the remaining 400 odd pages of the novel are the harrowing account of K.'s struggle to establish contact with the castle to clarify why he was summoned and what his role there is. K. seeks answers but he never succeeds. He never gets an answer.

Kafka's novel is not an easy read. I first read the book in my early twenties and have now, in the process of writing this book, picked it up again. It is one of those reads that is a slog, but I would never want to be without it. I believe that Kafka communicates something essential about our existence and our relationship to the world and he does it in a style that I find both elegant and utterly painful. With Kafka, the existential struggle of his main characters is a never-ending ordeal, page after page. It's like driving a U-Haul truck on highway 1 from Vancouver to Toronto — something I once did with a friend — it just goes on and on and on. But once you have read the book it stays with you. It is like ingesting an archetype, it becomes a permanent imprint on your mind, it establishes a new set of internal stars, which you'll be navigating henceforth. I find most of Kafka's writing is like that, timeless, painful, and of urgent importance.

But let's get back to the story. K. arrives at the village at night without knowing what his assignment there is and, so it seems, without knowing anything about the village or its customs. We don't know anything about his background either, not even his name. He is completely new in this world; the village is unchartered territory for him.

In the village, K. is assigned a liaison with the castle, a bureaucrat named Klamm, who reports to the council chairman of the castle. But K. never actually meets Klamm, nor does he meet any other official from the castle; they all remain distant, almost mythical figures throughout the novel.

Instead of an assignment at the castle, K. is offered a job as a teacher at the village school. This does not satisfy him, and he is determined to reach either Klamm or the castle to get answers concerning his presence in the village. He wants to know why he was summoned and what his role there is, and he refuses to make do with the village life that he is being offered.

K. soon finds himself in an endless maze of obstructions, divergences, and everyday trivialities that prevent him from getting anywhere near the castle. The closest he gets is a telephone connection, which only offers unintelligible noise that resembles distant voices or someone laughing. But K. never gives up, nor is he ever angered. With an oddly indifferent persistence, he counters every setback and failure with renewed efforts to reach the castle.

Much like Kafka's other novels, the atmosphere in 'The Castle' is one of estrangement, alienation and an eerie acceptance of the absurd. There is something incredibly *normal* about the endless chain of absurdities and grotesque dialogues that fill the pages of this novel, and which in the end add up to nothing at all. The village is like the swamp of everyday life, a swamp that K. tries to escape by establishing contact with the castle.

Kafka never finished the novel. Before his death, he gave the manuscript to his friend Max Brod with instructions to end the novel by having K. die in the village. In this version, K. was to receive a notification from the castle on his death bed informing him that his *"legal claim to live in the village was not valid, yet, taking certain auxiliary circumstances into account, he was permitted to live and work there."*

Like many other readers of Kafka, I like to think of this novel as a metaphor for our search for truth and as a metaphor for our search for scientific truth. For me, the castle represents the Truth, Nature itself, or even God, and K. is our representative, the surveyor, the seeker of truth. K.'s very presence is an accident, something that was not supposed to happen, an anomaly. There was never meant to be a surveyor in this place.

I believe the determination of K. is a key element in the novel. His hope and desire to reach the castle never die. Our thirst for truth is unquenchable and unstoppable and yet we, like K., accept the circumstances we find ourselves in and merely find them slightly odd. All of Kafka's writings operate in this existential limbo, where the absurd conditions that reality offers us are met with a merely slightly embarrassed look of discomfort.

It is interesting that in the Czech language, which Kafka spoke, the word Klamm means "illusion". Klamm is K.'s connection to the castle, a connection that we infer is not real. K. never meets Klamm — Kafka seems to tell us that our connection to ultimate truth is of a very fragile and illusory nature.

Another interesting detail is that the word 'Schloss' has a double meaning in German. Besides 'castle' it also means 'lock', which suggests that for K. the castle is locked, he will never get there, he will never open it.

So, K. is an anomaly in this world, asking questions that should not be asked and striving towards an unobtainable goal.

Perhaps I'm stretching the analogy when I think of Klamm as a metaphor for a final theory — the closest we will ever get to ultimate truth through our intellects, our liaison to truth, to God if you like — as Kafka was no physicist and, as far as I know, knew nothing of modern physics. But a symbolic text like Kafka's has its own life, which goes beyond the intent of its author, and thus I indulge myself: Klamm is a metaphor for the final theory, which is an illusion, literally. And K. represents we physicists and scientists, the seekers of truth, who never give up hope of finding the absolute. We soldier on, generation after generation, knowing well that most likely our effort is in vain. No matter how hard we try, we're never going to meet Klamm, we will never reach the castle — and we will never give up.

What does Kafka's book tell us about our scientific search for the ultimate laws of Nature? When I read this novel I wonder whether we will share the fate of K. and chase an illusory final theory through an infinite maze without ever reaching it? Whether the idea of ultimate scientific truth is an illusion and that our very questions, our presence as 'land surveyors' in this village of ours, is a mistake?

Perhaps this is the most likely outcome, the safe bet: that we will never find the theory. That we will never know if it even exists but will keep looking until we die; generation after generation, until some of our descendants in a distant future grow sick of it, or our civilization destroys itself. Perhaps. But before we accept this bleak outcome I suggest we watch a sci-fi movie.

CHAPTER 13.

SHELL BEACH

"Riding in circles, thinking in circles. There is no way out ... "

— Detective Walenski, Dark City

One of my favorite sci-fi movies is the 1998 film 'Dark City', written and directed by Alex Proyas. This movie has, just like Kafka's book, something archetypal about it. It touches the same general theme but reaches a decisively different conclusion. I think that we should take a break and watch it.

Dark City is dark. The sun never shines. The entire movie plays out in this city, where nothing seems solid, where everything is in flux. The city itself keeps changing, as do the memories and identities of its inhabitants. And nobody seems to notice.

In the first scene of the movie, we watch the protagonist, a young man, wake up in a bathtub not knowing who or where he is. He soon learns that his name is John Murdoch and that he is wanted by the police as a murder suspect. A serial murderer of women.

Murdoch, who suffers from amnesia, soon realizes that there is something very peculiar about the city he finds himself in. He sees what nobody else sees, namely that nothing stays the same, that reality is constantly being reshaped. The reshaping always happens at midnight, at a moment when the city clocks freeze, and everyone falls into a deep coma. John Murdoch is the only one who stays awake and sees the reshaping.

Dark City is an action movie — and like all good sci-fi movies, it has aliens in it, too. The aliens have the problem that they do not possess individual identities or souls, and for that reason, their race is on the brink of extinction. To prevent this from happening they perform experiments on the inhabitants of the city to learn the secret of individuality and to eventually acquire it. The key ingredient is memory.

An important character in the movie is the retired police detective, Walenski. He has lost his mind and has plastered his office with unintelligible notes (a sure sign in a movie that someone has gone mad). The picture that he keeps drawing is a spiral, the same image that is cut into the skin of all the murdered women. When Walenski is asked by his former colleague what he is up to, he replies that he has been

- *"riding in circles, thinking in circles, there is no way out ..."*

and later in the movie, detective Walenski meets Murdoch and tells him that

- *"There is no way out, you know. You can't get out of the city, believe me, I have tried."*

Their brief meeting ends with a

"... but that's okay, I have figured a way out"

---

whereupon Walenski jumps out in front of a train and kills himself.

None of the memories and identities of the people in the city are solid. That is what both Murdoch and Walenski have realized, they have awoken during the reshaping process and seen it, seen how everything is constantly changing. So, what should Murdoch believe then? What can he trust? Where can he stand? If everything is in constant flux how is he to orient himself?

In one of the first scenes, John Murdoch finds a postcard in his suitcase with a picture of Shell Beach, the place of one of his childhood memories. During the movie, this place becomes Murdoch's fixpoint. He wants to find Shell Beach to see ... what? To see if it exists, if there is something real there that he can hold on to.

He is searching for a foundation. A place to stand on and from where he can move forward.

So, in "Dark City" we find the same theme as in Kafka's "The Castle". Our protagonist is caught in a strange place, a maze-like town or city, and is unable to reach the place that could provide him with answers. The search for truth is what Kafka and Proyas are dealing with.

I find it interesting that both K. and John Murdoch lack a clear identity. K. does not have a name and John Murdoch suffers from amnesia, so they both want to find out who they are. K. wants to know what his assignment is, why he was summoned to the castle, and John Murdoch wants to find out if his childhood memories are real or implanted — he wants to find out who he is.

What I find so fascinating about Dark City is that Proyas reaches a different conclusion than Kafka does. Both K. and Walenski are caught in their respective mazes. They have tried everything but found no way out. Their exit is their deaths, the final door out. K. eventually dies and Walenski jumps out in front of a train. But with John Murdoch it is a different story, he eventually finds Shell Beach. He actually reaches the point where the answers to his questions can be found.

He finds a way out.

And what does Murdoch find? This is what interests me. He finds absolutely nothing. Shell Beach is completely empty. There is no beach, there is no ocean, there is no sky, it is a black void. There is nothing there at all.

John Murdoch has to create Shell Beach himself — and luckily, he has acquired the superpowers to do just that. The movie ends with him and his wife on the beach. In daylight.

I like this movie not only because it's fun and beautiful but mostly because I see it as a fascinating metaphor for our search for truth. I believe that this movie says something about what our scientific quest could have in store for us. Proyas shows us that the answer that awaits us may turn out to be different than what we are hoping for. That our search for truth may lead us to an empty answer.

Indeed, is that not the only imaginable outcome? That what we are looking for will turn out to be essentially empty, if we ever find it. Think about it. If the answer was not empty, wouldn't it simply open the door to yet another round of scientific reduction? If a theory involves non-trivial structures will it not automatically invite further questions and a search for yet a deeper layer, that can explain these structures? Isn't the only imaginable *final* theory one that involves only structures that are completely trivial and self-evident and therefore, essentially, empty?

I believe that Shell Beach is the most likely outcome.

CHAPTER 14.

WALKING UP THE BACK
OF A TIGER

Ignoramus et ignorabimus

— Latin maxim

Ry, Denmark, 1988

It is dark, I can hear someone walking around upstairs, maybe my dad, but otherwise, the house is silent. In the faint light from the window I can see my desk with the typewriter and above it the bookshelf that my father put up. There are my schoolbooks and the book on philosophy that I got for my birthday.

I'm 17 years old and I'm wide awake.

I'm thinking about the Universe. Is it finite or is it infinite? If it is finite, then what is outside of it? And what is *all this,* everything? And what are *we* doing here? Why are we here? What will happen to us when we die? What is consciousness?

Questions. My head is full of questions. And they all boil down to a single word: 'WHY'

It fills my body, the word WHY. It is an intense, overwhelming feeling of wonder, painful wonder, like a tension throughout my body. The word is pumping through my veins, it is vibrating throughout my entire body: why! why! WHY!!?

This is what keeps me awake: there MUST be an answer. Of course, there is. These are questions that must have a definite answer, either yes or no; up or down; left or right. This reality has a definite structure. It is *real*, and these questions must have an answer.

It must be possible — at least in theory — to *see* all this from the outside, to observe it, to understand it, to *know the answer*. That is: there *must* be an answer. I can feel it, the answer, that it exists, I can feel it.

And I want to *know* it.

But I never will. Nobody will. I know this too. It is obvious, and it pisses me off. There is an answer out there, somewhere — there *has got to be*! — of course, there is, but we will never find it. I am sure of it. It is not within our reach.

This makes me upset. It makes me want to climb out of this life in protest, file a complaint, and just quit. I can feel my entire body rebel against this injustice. I can't believe that this is the reality that I find myself in. It's just not fair.

What is the point of this? Why am I here in this reality, cut off from an answer but with a thinking brain capable of asking for it?

It's silent anger, I'm angry at a God I no longer believe in. I want to know the answer, I demand it!

Nothing happens. My room remains dark, the house remains quiet. The faint glow of the moon shines upon my typewriter on the desk. When I listen carefully, I can hear my brother breathing in the room next door.

I lie awake for a long time listening to the night. And to all the other nights to come after this one.

✡

Around 2007 we begin to tour the world, Johannes and me. I form the vanguard while he provides artillery support from the rear.

It is clear to me what we need to do. Neither of us has a permanent position, so we need support to be able to pull this off, to survive in the academic arena with this project. And we won't be able to convince most people of our ideas, they are too complex for that. They involve too much mathematics and different research fields that are too far apart. Very few people have detailed knowledge of both canonical quantum gravity and noncommutative geometry and very few physicists have the mathematical background necessary to understand the strategy that Johannes has come up with. And besides, our ideas are just that, *ideas*. We haven't developed them very far yet, so if we want to be able to sell this, we need to find the right buyers.

So, what we need to do is to seek out the best people in each field, the strongest and most visionary. That is the only way I see forward. And we need to focus on those communities that are directly related to our work, which means noncommutative geometry and loop quantum gravity.

✡

Potsdam, Germany, 5. November 2008

I meet Johannes at the hotel in Potsdam outside of Berlin. He arrived a few hours before me and I find him in the hotel restaurant.

\-   *"Hey, how are you doing?"*

I sit down next to him. He's on his computer and I can see that he's reading some physics paper. He looks up and greets me.

\-   *"What are you reading?"* I ask.

\-   *"Just some of their papers, I wanted to check out some details of what they are working on".*

We're here to visit one of the leading physicists in loop quantum gravity. I contacted him about a month ago — a young German professor, very influential in his community — and he arranged for us to visit. I've never talked to him before so I'm a little anxious. I've seen him at conferences a few times and he comes across as a cross-section between a physicist and an elite soldier. But I know that he's highly respected and I've been wanting to meet him.

The next day we take the bus to the university and find our way to the institute for theoretical physics. The young professor is on the second floor and we find him easily.

I'm slightly introverted and I always find these situations a little trying: to meet someone for the first time to discuss physics. They're such delicate things, ideas. They're personal, sensitive too, and discussing them with people you don't know is a little like showing them your most treasured secret and hoping they will be gentle with you.

I knock on the door to his office and he immediately responds. After a brief greeting, he says that we should go downstairs to the meeting room. I'm not certain what he means but I guess that we're going to get a cup of coffee and start with a little small talk. It's Monday morning and we've got three days, so there is no need to rush things.

As we walk through the building we meet a few younger people who appear to be expecting us — we're introduced and shake hands — and as we arrive at the meeting room, we've increased to a little delegation of perhaps eight people. When we walk into the room, I see about ten more people already seated, all facing the large blackboard near the door.

The young German professor and the others all sit down and look at Johannes and me, who remain standing by the door where we feel increasingly awkward. The professor then nods at the blackboard with a "please".

This is one of those situations you sometimes find yourself in where jumping out the window doesn't seem like such a bad option.

I quickly look at Johannes — he looks back with a bewildered expression — and since I know that it's going to be me who has to deal with this situation — after all, I was the one who organized this visit — I pick up a piece of chalk and walk towards the blackboard.

I don't really know what to do but I quickly improvise a blackboard presentation that ends up lasting almost three hours with no breaks. The audience sometimes asks questions, but mostly I'm just firing all my canons in random order and hoping that they'll at least agree to take a break for lunch.

We spend three days in Potsdam and each day Johannes and I take turns at the blackboard. Tuesday Johannes talks about noncommutative geometry, mathematical background material, and I mostly just sit and listen and occasionally take a look at the audience and wonder about this odd visit, where an expected discussion between 3-4 people has turned into an improvised marathon lecture.

At one point I glance across the room where the young professor is sitting — the tables are organized in a horseshoe and I sit on one side of the room directly opposite him — and I realize that he is not paying attention to Johannes' presentation but is instead staring directly at me with a strangely intense expression. It feels incredibly awkward and I quickly turn my head and look at the blackboard. But a little while later I can't stand the itchy feeling of his stare and once more, I look across the room. He is still wearing that same expression, as if he's looking at some peculiar creature he has never seen before, trying to determine what to think of it. It makes me so nervous that I abruptly stand up and, bewildered, ask my neighbor where I can find a toilet before walking out of the room.

One of the key issues that distinguishes loop quantum gravity from our approach is a choice of graphs — and by 'graph' I mean a selection of points with edges that connect them.

Graphs play a key role in their theory, where both the algebra and the construction of the Hilbert space rely on choosing an ensemble of graphs as large as one can imagine. And what we're telling this group in Potsdam is that if you want to employ the mathematics of noncommutative geometry, then you'll have to work with a smaller ensemble of graphs, cubic graphs for example.

This is controversial and it's clear that nobody in the room believes that what we're saying makes sense. By choosing a smaller ensemble of graphs you're almost certainly forced to change the way you view the graphs — from being a quantity with physical significance to being merely a computational device — and thereby the entire strategy behind the theory. Nobody in loop quantum gravity appears willing to take that idea seriously. We're breaking doctrine here, the choice of graphs is sacred ground in the loop quantum gravity community, and to suggest any changes appears to be heresy.

It's a strange experience — this tour of ours through the different communities of modern theoretical physics. Each community has a set of fixed beliefs, which, when challenged, will call upon the perpetrator the full wrath of all the Gods — or so it seems — and yet at the same time, everyone is more than happy to challenge anything that comes out of the competing communities. It is becoming increasingly clear to me that to be accepted by a community you must adhere to their core beliefs — and by not doing so you're placing yourself outside their village walls, you're an outsider, you're noise.

On Wednesday afternoon, after three full days of lecturing, we finally seem to be done. As we're about to wrap everything up, Johannes and I ask the young German professor if we could perhaps also ask him a couple of questions about his work. He agrees and stands as if ready for combat. I then ask him a question about some papers he recently authored. He answers my question with a single "yes" and looks at me with a blank expression. After an awkward pause Johannes then asks another question — some mathematical detail — and this time the answer is "no".

The professor then asks if that was all and as Johannes and I just stand there unable to think of anything intelligible to say he then shakes our hands and walks out of the room. That is the end of our visit.

✡

Mumbai, India, 21. February 2011

I have given lectures in so many countries — in Canada and the USA, in most of Europe, in Australia, and even in Iran. And now I am in India to give a lecture at the Tata Institute in Mumbai.

One of my friends, who is a Ph.D. student in string theory, is studying here for a semester and I thought that I'd arrange a visit. I have given so many talks for experts in either loop quantum gravity or non-commutative geometry — or some related field — but the group in

Mumbai mainly works in string theory and I thought it could be interesting to try something new. To give a talk to a group of string theorists.

I like India very much. I have traveled in the country several times, mostly in the northern parts, and I like the atmosphere, the energy, the colors, sounds and all the strange sensations. And Mumbai is like the rest of India: huge, noisy and incredibly interesting.

My friend has told me that I should be prepared for a lot of questions during my seminar. The talks here often drag out due to all the questions, he said, sometimes for an hour and a half and even longer.

I'm a little nervous, and on top of that I've got solid jetlag. I am used to talking to audiences who are familiar with at least one of the two research fields that I talk about: loop quantum gravity and non-commutative geometry. But most physicists here will probably not know a lot about these topics and even worse, if they are just a little like the other string theoreticians that I have met, they will have a strong dislike for loop quantum gravity.

I arrive at the institute about an hour before my talk and after a few introductions, I finally stand in front of an audience of perhaps twenty-five people in a small seminar room. Among them is a very famous Indian physicist. It is to him that I'm mostly interested in talking.

I have prepared a fairly long talk. I begin with a short overview of non-commutative geometry — the key ideas and results — and after a couple of minutes, the famous Indian physicist raises his hand and interrupts me.

"What is a C*-algebra?"

he asks.

A C*-algebra is one of the key mathematical ingredients in quantum mechanics, it is a fundamental concept, and the fact that he chooses to ask me precisely this question confuses me. I know that he has published a considerable number of papers related to non-commutative geometry — about ten years ago a certain version of non-commutative geometry became fashionable within string theory — so surely, he must be familiar with this concept?

I explain to him what it is and then I continue my presentation. The audience listens in silence and I begin to feel ill at ease. After a few minutes, the famous Indian physicist stands up and leaves the room, and after that my talk feels like a long and lonely walk in the desert. I have lost them, there are no more questions, they have no idea what I am talking about and the only reason why they all remain seated is that they are too polite to stand up and leave.

I wish I had prepared a shorter talk — towards the end, I skip a few slides and half of my conclusion. I finish with *"thank you"* and deliberately omit to ask if there are any questions. I know there aren't.

Vatna Jökull Glacier, Iceland, March 2006

I can hear Niels digging outside the tent. I just woke up and now I'm debating with myself whether or not I should get out and help him.

- *"Niels, do you need help?"* I shout.

I'm hoping he'll say no. I can hear the wind howling and have no desire to go out there.

- *"Hey man! Yes, get out here, the girls' tent is almost completely gone, we need to dig them out!"*

I turn on my headlamp and get out of my sleeping bag. Before I put on my gear, I take a look at my heels. I made the mistake of bringing new ski boots instead of my old ones and I'm paying the price. My heels are hurting badly.

I pull off my sock and peel up the edge of the band-aid to look at the wound. I immediately realize that this is a mistake: all the skin of my heel peels right off and exposes an ugly sight of bloody flesh. I quickly put the band-aid back on, get out some tape, and secure the whole thing with several loops around my ankle. I'm not going to reopen this package before the end of the trip. Then I get dressed, put on my boots, and crawl out of the tent into the snowstorm where I start digging with Niels.

We're on the very top of the large Vatna Jökull glacier It's March and we're hunkered down in a minor storm. There are six of us: three men — all foreigners — and three women — all Icelandic — and we've set out to cross the glacier on skis, a trip that we expect will last between a week and ten days — depending strongly on the weather conditions. There are three possible exit points for us to get off the glacier, one of which involves a long abseil, and we are still debating which exit point to aim for.

This is a trip I've been wanting to do for a long time. The Vatna Jökull is one of the most beautiful glaciers I have ever seen, enormous too, stretching almost 150 km across the southern part of Iceland and reaching a height of 2100 meters above sea level. The best time to cross it is in deep winter when all crevasses are covered with many meters of snow and you're relatively certain not to be hit by rainstorms.

Getting to the western edge of the glacier where we started the trip was a major undertaking. We had contacts in the Icelandic civil defense who provided two super jeeps — large Toyota 4WDs with modified chassis that hold enormous wheels the size of a grown man — to drive us to the rim of the glacier, a place you can otherwise only access by air at this time of year. It was an eerie experience to drive through the deep snow seated in a comfortably warm cabin, drinking hot tea and listening to music, to then step out in -15 degrees Celsius plus wind chill with all our gear and watch the jeeps speed off in a cloud of white powder.

During the day we walk in a long line, each following the tracks of those ahead and pulling a sled with equipment. When the visibility is low it is virtually impossible for the one in front to walk a straight line — you constantly think you see a fixpoint but all fixpoints are imagined and always on the move, with you walking in arches trying to follow them — so we have the person in the back walk with a compass and yell commands: *right! left! right!* which we then repeat up through the line to help the dizzy friend in front to stay on course. It's a comical situation, a bunch of people in a soup of white screaming at the poor person in front, who is losing their mind in frustration.

It's a terrifying place. I have the feeling that we're ants walking up the back of a sleeping tiger. *Please, don't wake up, dear tiger.* I know that if we're hit by a major Atlantic storm this could quickly turn into a very serious situation.

But the tiger is sleeping, and we've enjoyed some incredible scenery so far. There is a persistent weather pattern coming in from the north, which means very low temperatures but most importantly, stability.

It's good to be back in Iceland and it's good to see Niels and my other friends again. But skiing across Vatna Jökull doesn't invite conversation: the intense cold and the physical exertion makes everyone focus on the immediate task of pulling their sled. This is precisely what I came for, to empty my mind and just forget everything for a few days. When you're in the snow, what matters are your gloves, melting snow, navigating, eating, nursing your feet and securing your tent. Thinking about mathematics is a luxury that I can't afford here and that is exactly how I want it.

CHAPTER 15.

NONCOMMUTATIVE GEOMETRY

"Why is a raven like a writing-desk?"
"Have you guessed the riddle yet?"
the Hatter said, turning to Alice again.
"No, I give it up," Alice replied: "What's the answer?"
"I haven't the slightest idea," said the Hatter.

— Lewis Carroll,
Alice in Wonderland

Out of the trenches, we are now running through a landscape, unlike anything we have seen before.

We have come to what I believe is the most interesting research field in contemporary mathematical and theoretical physics. Noncommutative geometry.

As the name suggests, this is about geometry. And since geometry is synonymous with general relativity, we're also talking about Einstein's theory.

Noncommutative geometry begins with a new mathematical formulation of the geometry of spaces, which in a sense turns standard notions of geometry upside down. This new formulation automatically leads us to a generalization of geometry that will not only take us far beyond our usual notions of what a geometrical space is, but also take our breath away as it offers a new stunning view of the landscape ahead of us.

But before we throw ourselves into the thick of it, let me give you a friendly warning. Noncommutative geometry is a highly advanced research topic. It involves concepts that are not easily accessible, even to experts. I will do my very best to clear a good path for you, but you might still find the terrain rough. Don't worry! You're in good company. Many people in modern physics are confused about Connes' work — but this topic is important, we must pass this way, so hang on and follow me!

Let us first look at how we measure distance by starting with something very simple, a straight line on a piece of paper. If we want to measure the distance between points A and B on this line, what should we do? Well, we just get out a good old ruler, place it on the paper and read off the distance. No surprises here.

But let us try something different. Let us consider instead functions on the line. A function is simply an assignment of numbers to each point — like different temperatures on a timeline. Since we can pick different functions that we can add and multiply to obtain new functions, we are really dealing with an *algebra* of functions. So, we choose one function and take the difference between the values that this function has at point A and point B. That is, we subtract the value that the function has at point B from its value at point A.

"Well, but this could give us anything!" you may protest. Yes, you are of course right, so for this to make sense we need some additional information. If, for instance, we knew that the chosen function has a rate-of-change that equals exactly one, i.e., that its gradient is one, then the difference between its values at B and A will equal the distance between A and B. Voila!

This may seem completely trivial to you, and you are right, it is, but just stay with me a little longer.

So, we have two different ways to measure the distance between our two points. First, we have the points themselves and a ruler, and second, we have functions on the line and information about their gradient. In the first case, we use the ruler to tell us the distance and in the second case, we subtract the values that the function has at the two points.

Let us see what this amounts to when we have something more involved than just a straight line on a piece of paper — let's consider a three-dimensional space. In the first case, where we had a piece of paper and a ruler, we have now the three-dimensional space, i.e., *points*, and a metric field that tells us how far apart these points are. The metric field is our ruler and the combination of 'space + metric' gives us a geometry. This is what we call Riemannian geometry. The basis of general relativity.

But what about the second case? It turns out that the second option can be generalized to something that is far from trivial. In this case, we have replaced the space, i.e., the points, with the *algebra of functions* on this space. That is, we have the collection of all functions on the space. And to measure the rate-of-change of the functions — which in our example with the straight line was necessary in order to measure distances — we have a certain rate-of-change or *gradient* operator called a Dirac operator. These two elements, the algebra of functions and the Dirac operator, need a place to live, which is a Hilbert space: our third ingredient.

So, in the second case, we have a collection of three mathematical entities: an algebra, a Dirac operator, and a Hilbert space. This is what is called a *spectral triple*.

Notice that the second option involves the language of quantum mechanics: we have operators, algebras, and Hilbert spaces. The reason for this is that the entity that tells us the rate-of-change of a function is an operator and therefore needs a Hilbert space to live in.

What makes spectral triples interesting is the following two crucial facts. First, it was proven in 2008 by Alain Connes that the ordinary formulation of a geometry in terms of a space and a metric — the first case with Riemannian geometry — is *completely* equivalent to a formulation in terms of a spectral triple.

It does not matter which formulation you use; they amount to exactly the same. This implies that we have found an alternative formulation of general relativity[3].

Also note that the space itself – the *points* – is *not* a primary ingredient in the spectral triple formulation. The algebra is at first an abstract object, the space with its points emerges as a secondary ingredient.

---

[3] There are some important details that I could mention here, but I want to make the going as easy as possible so we'll leave them for some other day!

Here is the second fact, that makes spectral triples interesting: it turns out that the mathematical rules, that a spectral triple is required to satisfy, are straightforwardly generalized to apply to a much larger class of algebras known as noncommutative algebras — hence the name noncommutative geometry.

This means that this *equivalent* formulation of Riemannian geometry and general relativity opens the door to a whole new landscape of geometries, exotic geometries that we did not have access to with the ordinary formulation.

But before we go through that door let us pause for a moment and recall what a *noncommutative* algebra is.

When we discussed the second method of measuring the distance between two points on a straight line, we didn't pause to consider whether the functions that we used commute, i.e., whether it matters in which order we might multiply their values at different points. And of course, we didn't, these values are just numbers, it does not matter in which order they are written.

But remember our discussion in chapter four where we discussed quantum mechanics. There we encountered operators that did not commute, and we learned that these operators form algebras, which are the key ingredient in quantum mechanics. A central point in that discussion was that these algebras are *not* commutative. They are what we call noncommutative. It matters in which order we multiply the elements in these algebras.

And now we learn that there exists a natural generalization of Riemannian geometry and general relativity that involves such algebras?

If this makes you a little dizzy, that's completely appropriate. And I assure you, the craziest stuff is yet to come.

✡

The great pioneer of noncommutative geometry is the French mathematician and Field medalist Alain Connes.

The first time I saw Connes was in Vienna around 1999. He gave a talk at the Schrödinger Institute that year when I was a student at the Technical University just a few kilometers away. I knew some people at the Schrödinger institute, and they had told me to come over, *you've got to see this guy*, they had said.

I did not understand much of what Connes said that day. I was not familiar with the math. The words, or the concepts. This was a completely new landscape for me, a new continent. Yes, it was a different galaxy actually. But I understood his gestures, his way of presenting it all. I got that part.

What I saw was a man on fire. A tall thin man with a beard and white hair that stood out from his head like a halo in wild disarray. When he talked, his shirt came untucked. It left his pants and started flying around him like a white shadow that had trouble keeping up with his erratic movements and just barely managed to hang on to his body.

Connes was jumping, he was running, he whirled around, he laughed while his arms went flying up and down in all directions. I witnessed a modern dance piece in the guise of a math talk. If you didn't know that he is one of the greatest mathematicians alive you might think you were watching a madman.

But it was his face that I remembered after his lecture. It was the face of a child. A child that has seen the light, a child that has just received the Christmas present of his life, exactly what he had wished for just much, much better. And now this child was telling all his playmates about his wonderful present; his face engulfed in pure joy.

I couldn't help being infected by the joy of this child, by the energy of this man. To be drawn in. Into Connes' sandbox of wonders.

✡

One thing we should be clear about when we talk about spectral triples is that although this new formulation of Riemannian geometry involves the *language* of quantum mechanics, i.e., operator algebras and Hilbert spaces, this does not mean that it *is* quantum mechanics. It is not. This is geometry and it is classical.

It has the potential to involve quantum theories. At least that is what we think.

But now we are ready to walk through that door I talked about. The door into the world of noncommutative geometry.

So, we are going to consider spectral triples that involve a noncommutative algebra. And since the algebras that we have talked about so far are simply algebras that correspond to ordinary spaces — general relativity and all that —it might be good to start with something similar.

Let us consider an algebra that is a combination of an algebra of functions on a four-dimensional space (which we know corresponds to ordinary geometry) with something else. And this something else is going to be what we call a matrix algebra.

What is a matrix?

A matrix is a mathematical object that encodes, for instance, a rotation in a space around some axis. It is an array of numbers, three by three for example if the space is three-dimensional, and there are rules, which tell us how to add and multiply these matrices.

And then you can form an algebra out of the matrices.

Now, if we combine our algebra of functions over an ordinary space with a matrix algebra we get something that is called *an almost-commutative algebra*. This is an algebra that is almost identical to the commutative algebras that correspond to ordinary geometry, we have just added some matrices.

And here is a surprising fact: If you pick the right almost-commutative algebra and build a spectral triple over it, then instead of just general relativity you get general relativity *plus* the standard model of particle physics.

Wow!

Let us walk through that statement again.

Remember, a spectral triple is an alternative formulation of ordinary geometry, where we use algebras, Hilbert spaces and this gradient operator called a Dirac operator instead of just a space with a metric field. And spectral triples have the magical feature that they also make sense when the algebra is not just of the ordinary type that corresponds to the spaces that we are used to, but also more exotic algebras that involve noncommuting elements.

And then the first example we come across, where the algebra has just been slightly changed, gives us the standard model of particle physics coupled to general relativity?

What does this mean?

Well, I think that the honest answer to that question is that we don't really know yet. But let's take a closer look at what we do know.

First, this clearly shows that it is possible to understand the whole thing, Einstein's theory of general relativity plus the standard model of particle physics, as a *single* gravitational theory. *One* theory of general relativity over a strange kind of space.

Secondly, this new formulation changes the way we see the standard model. There are a number of details in the standard model, that are cast in a completely new light within this new formulation. Take for instance the Higgs particle that provides all the masses in the standard model and stands out as the only one of its kind (it is what is called a *scalar* field, as opposed to the other fields, which are either *spinors* or *gauge* fields). In the spectral triple formulation, the Higgs particle and the gauge fields emerge as two aspects of one and the same thing, a kind of unified entity that makes perfect sense within the framework of noncommutative geometry. Without Connes' new mathematical machinery we would never have seen this.

The whole standard model is rewritten here in a language of geometry. It is very beautiful. Very odd too. It blows my mind.

✡

The spectral triple formulation of the standard model of particle physics is the product of a long collaboration between Alain Connes and the particle physicist Ali Chamseddine. The two have formed a team in theoretical physics over the past 20 years or so, and they reinvigorate a tradition, which previously characterized theoretical physics: the intertwinement of theoretical physics and mathematics.

Quantum mechanics and Einstein's theory of relativity emerged as a product of both research fields, physics and mathematics. These theories took their first steps on both of these legs and many of the heroes, whose names any physics student knows by heart, were mathematicians. Dirac, Weyl, Hilbert, von Neumann, Wigner, Birkhoff, Wiener, Jordan, Klein, Gordon, Cartan, etc.

The connection is still there, it probably always will be, but it is not in its prime any longer. I know many physicists who view mathematics and mathematicians with skepticism. I may be too young to fully understand this, but I suspect it is the inability of mathematicians to make sense of quantum field theory that has caused this. *What good are they if they cannot figure that one out?*

But Connes and Chamseddine are an exception. I first met Chamseddine when I was a student at the ETH in Zürich. I had completed my bachelor at the University of Aarhus in Denmark and my move to Switzerland was motivated in parts by a wish to submerge myself in theoretical physics and in part by my passion for whitewater kayaking — an activity that is completely incompatible with Danish topography.

The University of Aarhus is a small yet very good university, but its focus is experimental physics. Everything points in that direction and if your inclinations are in the theoretical you will either have to stay in the closet or move. I moved and it was heaven. At the ETH in Zürich, I found all the theoretical physics courses that I could ever dream of, particle physics, axiomatic quantum field theory, Riemannian geometry and general relativity, the standard model, and many other topics. And Chamseddine was one of my teachers, in a class on particle physics.

He is a tall man, with dark hair and a mustache. I was too shy to talk to him back then but later I got to know him a little. He comes from Lebanon where he is a professor at the American University in Beirut.

What Chamseddine and Connes are doing is a kind of experimental mathematics. We know experimental physics, which is the field of research where experiments are conducted to get clues as to how physical theories should be built. Chamseddine and Connes do something similar: they take the standard model of particle physics as their experimental input and ask what new mathematics lies hidden in its structures. This is how the research field of noncommutative geometry came about, by looking at Nature. And that is very much Connes' philosophy as far as I understand him: to view the standard model as a mathematical statement. *This is what Nature has given us, what is it trying to tell us?*

✡

Essentially the spectral triple formulation of the standard model is a *reformulation* of what we already know. There are important new relations, and the number of free parameters is smaller than in the ordinary formulation, but that's not the main point.

It reminds me of Maxwell's equations, which have two different formulations, one in terms of electric and magnetic fields and one in terms of a gauge field. These two formulations are *completely* equivalent. It makes no physical difference which one you work with. But the gauge field formulation points towards other, deeper physical theories, like Einstein's theory of relativity and the standard model. The point is that it is the formulation with gauge fields that best *captures the deep mathematical structure* of Maxwell's equations.

I think that this is how we should understand the work by Chamseddine and Connes. It is a road sign. It points towards something, which we do not yet understand.

What we need to do is to read this road sign very carefully.

Two key elements stand out. The first one I have already told you about. It is the *almost-commutative algebra* that took us from general relativity alone to general relativity coupled to the standard model.

The question is where that algebraic structure comes from? Why does Nature give us an almost-commutative algebra? That is the key question, the thing to remember: *the almost-commutative algebra.*

The second key element that stands out in the road sign is quite odd.

The thing is that what Chamseddine and Connes find is not exactly the complete standard model but rather its *classical* counterpart. The quantum aspect of it is not there. What they do is to add this part by hand afterward, they simply quantize the standard model using standard methods of quantum field theory while leaving out the gravity part. Gravity remains classical.

- But wouldn't you think that the quantum aspect, the quantization, should somehow be an integral part of this formulation if it were to be fundamental? You may ask.

Yes, well, I certainly agree, but how? The point is that the spectral triple formulation essentially deals with the geometry of space and time, it is general relativity extended to this odd matrix algebra, which forms the almost-commutative algebra. If there is one thing that we know about general relativity it is that the standard method of quantization runs into problems when we try to apply them to gravity.

This is remarkable. On the one hand, we have the old formulations, where gravity and the standard model are separate things and where the standard model is a quantum field theory.

And on the other hand, we now have this beautiful and surprising spectral triple formulation, where everything is geometry but where quantum field theory does not really fit in.

This raises the fundamental question: *what is quantum field theory?*

The point is that in Chamseddine and Connes' formulation *everything* is formulated as a *single* theory of gravity. Any quantization must, in this view, somehow be a quantization of geometry. If we are to take this seriously, then mustn't quantum field theory be understood as a low energy limit of a theory of quantum gravity?

And remember that we don't know exactly what quantum field theory is in terms of rigorous mathematics in the first place and that many people believe that the standard model really *is* a low energy limit of something else.

So perhaps we should understand Chamseddine's and Connes' work as a hint at where the standard model comes from, namely that it is the low energy limit of a theory of pure quantum gravity?

And that is the idea that first got me interested in loop quantum gravity: I believed that we should search for an explanation for the almost-commutative algebra found in the mathematics of the standard model in a setting of pure quantum gravity. And I believed that quantum field theory would eventually turn out to be the low energy limit of a theory of quantum gravity. This was the approach of my work with Johannes, this is where we started.

As it turned out, what we eventually found is very different from what we were looking for. But we will get to that later, we need to cover a little more ground before we are ready for that.

✡

Believe me, there are many details in this noncommutative geometry story that I am leaving out. I have to, this book would sink like a stone if I didn't. But there is a purpose besides buoyancy for leaving out all the nitty-gritty details, which is that although obviously essential, they can block our view of the landscape.

Sometimes when Chamseddine's and Connes' formulation of the standard model is discussed among physicists it ends up in a technical argument about whether this formulation involves more or fewer free parameters than the ordinary formulation and whether it has produced post- rather than predictions and other tedious nitpicking. This is the accountants speaking. This is the noise of details. What you should look at here is the big picture, the beauty, and where it might lead us. And there is beauty here, I assure you.

Of course, in the end, details are everything. Either you've got it right or it is wrong. Period. But until we have found the right direction to march off to, we need to keep an eye on the big picture, we need to think not only as engineers and bookkeepers but also as artists and adventurers.

We need to be *pirates*.

So, let's keep an eye on the horizon and onwards!

CHAPTER 16.

GOING OVER THE EDGE

"There is no authority
who decides what is a good idea"

— Richard P. Feynman

Oberwolfach, Germany,
10. September 2009

"Stop being nervous!"

Ryszard whispers to me and sends me an angry look.

I have difficulty sitting still. I get nervous before I give a talk. Always. The guy speaking before me is almost finished. Now he is asking if anyone has any questions. It will be my turn in about two minutes, and I am nervous.

We are around fifty people in the lecture hall at the Oberwolfach research center. It is the big annual meeting on noncommutative geometry that is held here every year in September, and I have been invited to present our work to the community. And to Connes. I have the feeling that this is Connes' meeting and that I am here because he wants to know what our work is about.

✡

When you kayak a big waterfall there is a moment in the river just before you go over the edge, where you can no longer escape the pull of the current but where you are still in relative safety, still in calm waters. It is the point of no return.

At this point you see the horizontal edge in front of you. You see the river disappear and you hear it, the roar. You know what is coming, you know that the relative calm you experience right now is an illusion, it will soon end, and you will enter a chaos of cascading water.

All you can do at this point is to hold your paddle tight, make a few strong strokes to get the horizontal speed necessary to avoid being sucked in at the bottom of the waterfall. Then you hold your paddle away from your head and your body, so that it won't smash your face or break your neck when you hit the surface below the fall.

If I had known beforehand what was at stake here, I would probably have been too nervous to talk.

But the talk goes well. I do as I always do, start with the motivation, describe the problem. I present the big picture, where are we now and where do we want to go? I paint the landscape, the clouds, the forest, all the little birds and the ocean behind the hills. This is what I like the most, talking about the ocean behind those hills. How wonderful it must be.

Connes is sitting in the front row to my left. My talk soon develops into a conversation between him and me. Or an interrogation, perhaps. Depending on how you see it. But I like him, his energy is incredibly intense.

Then I get to the technical stuff, all the graphs, the infinite system of approximations, a tower of Hilbert spaces, the Dirac operator. I always dread this part of my talk a little. All these details. Sometimes I have to talk to Johannes on the phone the evening before I give a talk to clarify something with him, something that I have forgotten. It's the math. I get it, I really do, but I forget it too. It's just not the way I think, I see the picture, the landscape, all these details are just little rocks that my brain is somehow unable to hold on to.

But today is different, it is as if I feed on Connes' energy, his attention, his light. I have the feeling that I am juggling my rocks, all of them. They are light as air. I barely need to touch them before they fly up like friendly butterflies at my command. At one point, Connes exclaims *"yes, of course!"* — that is when I explain the construction of the Dirac operator — it is so natural and he sees it too, of course, he does. I love him.

But he does not see it.

Afterward, when people have asked their questions and I have answered them the best I could, after Ryszard has stood up and given his version of what we are doing and where he thinks we are heading, after all that, I hear

Ryszard and Connes walk down a corridor away from the lecture hall arguing.

"But there is a spectral triple here!" I hear Ryszard exclaim.

"Yes, yes, sure, I know, it's nice, but I am just not convinced", Connes replies.

✡

I don't mind. I feel great. I always feel good after going down a waterfall, after having presented our work, our ideas. I feel that I have given all I have; the rest is not up to me. People must make up their minds themselves. I am just relieved that I didn't get sucked back into the fall or break my ribs. I'm alive and in one piece and that is all that matters.

While I wait in the queue by the coffee machine a French mathematical physicist walks up to me. I vaguely know him. I know his work well. I respect him a lot, and I like him.

"You are a very courageous person, you know. I want you to know that", he tells me.

I am a little taken aback by his comment. Flattered. I still feel the kick from having presented a talk to a room full of the world's leading experts on noncommutative geometry. I am flying, everything around me feels a little unreal.

But later, when I am back down on the ground, I begin to think about what the French mathematical physicist had said. Why did he say that? Of course, for me personally, it did take a lot of guts to give that talk, it always does. I fear audiences. I prefer to sit at the back of the room. I prefer writing emails instead of talking to people, so yes, it took courage. But why did he tell me that? They all give talks here, it is what everyone does, right?

But of course, I do know what he meant. He meant that I had the nerve to present my own ideas. That I talked about *our* visions, where *we* believe this field should go, how *we* believe that Chamseddine's and Connes' work should be interpreted. And all this comes from me, it comes from Johannes, some of it from Mario and Ryszard too, but *none* of it comes from Connes. *That* is what he meant.

That I talked about an approach that hasn't been *sanctioned.*

But is that not what scientists do? Isn't that our role in all this? To develop and present our own ideas?

Münster, Germany, 28.11. 2008

The train is late. It should have been here half an hour ago, but they have just announced on the loudspeaker that it is an hour late.

It is cold. Almost midnight. I am standing on platform 7 at Münster railway station waiting for the night train to Copenhagen. Going home.

I'm tired. I feel drained. It has been a long week. We have been working on our project, Johannes and I, trying to move forward, to finish the mathematical proof that we have a spectral triple. But there are many complications. Johannes keeps finding small details that need more attention. I find it difficult to help him in this kind of work, it's really beyond what I am capable of. With this proof we're in the domain of advanced mathematics where I'm more a student than a collaborator. But I try to keep up and ask questions, I know that good questions help him.

Research like this feels like walking in a huge maze with just a tiny flashlight that has almost run out of batteries. We walk down one corridor, then another. Sometimes it takes a year before we understand that a corridor that first seemed promising, leads us nowhere, and then we retrace our tracks and try another way. There are moments when we discover a hidden door in a corridor that we had already walked down and dismissed as a dead end and we go back and explore this too, until we once more believe it to be dead. Trial and error and press repeat, and always there is the possibility that we're in the wrong maze, that this is simply one gigantic dead end. While the clock is ticking.

I talked to some people at the math department in Münster, mathematical physicists, I tried to explain to them what we are doing, to convince them too. I did my best, I drew a lot of graphs on the blackboard, waved my arms a little like Connes does but without the halo, and hoped that they would see it. But they didn't see it, they just sat there and looked at me, they tried to follow but I lost them somewhere along the way. And now I feel so tired. It saps energy, all those details and never a clear picture. Johannes and I think we see it, but the truth is that we haven't actually seen it, the ocean. Not yet. We just see the hills.

But I know it is there. Don't ask me how I know this, but I do. Just behind those hills, trust me, it's there.

I can smell it.

What frustrates me is not that the ocean is so far away and out of sight. That is understandable, we're building a theory and that takes time, lots of time, you cannot expect to reach the goal this soon. You mustn't ask too much of yourself.

It is the fact that I seem completely unable to convince others of our ideas. That we have an idea. And that it is worth pursuing. There are so few ideas around, they come scarcely, everyone is hungry for a new direction, so why won't anyone join us? Most of them work on stuff that everyone knows will lead nowhere, or at least should know — some of them even admit as much to me themselves.

They told me that we are too ambitious. That was their response when they realized the scope of what we want to do. It's too much, don't reach for the stars, don't jump, you might fall. That's what they think. You're crazy. And that is what makes me so tired.

I have the feeling that I am fighting a huge mountain of marshmallows. No matter how hard and how many times

I punch it I am unable to leave a mark. It just bounces back and doesn't even notice.

I throw a few euros in a vending machine and get a chocolate bar for the wait. I put my bag on a bench, get out my woolen hat and scarf, and sit down on the bag so that I won't feel the cold from the wooden planks.

There is something else. It's a kind of aggression or anger. Some of the older physicists seem outright hostile, as if it is a personal offense when they hear someone present a new idea. This part I find very hard to handle, the aggression.

I thought it would be different, I expected people, especially the young physicist, to be open for something new and different, something with a taste of adventure. I thought that they were hungry for rebellion too, but instead people seem to think more about securing positions and playing it safe. Perhaps this problem of quantum gravity has simply been around for too long. Einstein tried too, everyone tried, and nobody found a solution. So now it's considered sacred territory, only fools and geniuses dare go there — and we are no geniuses.

But I think they are wrong — when Bohr, Heisenberg, Schrödinger and Pauli and all the other heroes discovered quantum mechanics, they were no geniuses either. No, they were mortals, just like most of us, clever people with different talents but they had one quality in common: *they were hungry*. And they tried. They gave it all they had. They believed that they had a shot at it, and they *wanted it*. They were not afraid.

Among all the physicists I have met — and I've met a lot by now, I've given talks everywhere, or so it seems — I haven't met more than a handful of people who were truly trying, who believed that *they* had a chance of solving this problem. And among the young people, I don't think I've met a single one.

Why aren't people hungrier? That is what I don't understand. Why is there so little rebellion? What is holding them back?

People look to the leaders of the various tribes for directions, but that is not where new ideas are born — revolutions don't come from the top, they come from the bottom, from the nobodies, who become the somebodies because they tried. But if the nobodies don't try, then ...

I'm getting cold. I get up and walk over to the vending machine and throw in some more money. A Coca-Cola falls into the tray.

Time is running out, too. I've got a few more years and then what? Another grant? It seems almost impossible. You can only push a career so far on short research grants. At some point, you must reach the shore, or you drown. But I will never land a permanent position without convincing people of our ideas. The communities must accept us, otherwise, I won't get hired.

And the craziest thing of it all is that I don't really *want* to get hired, I'm not meant to be a university professor, it's not who I am. What I want is to find someone to *help us*. To join us. I just want to pass this idea on to someone else. It's too good to just let it die.

Too many thoughts, I wish the train would arrive so that I can get a few hours of sleep before Copenhagen.

✡

I know what the real problem is. It's the war. It's raging around us, the bombs are falling *as we speak*, the roads are mined, the frontline is here, right here where I stand, I can see the trenches of both armies, to my right and my left, filled with stormtroopers ready to go.

I didn't realize this at first. I heard comments, *"these people have no scientific credibility whatsoever"* and *"I can't be bothered reading that stuff!"* but I didn't pay attention until the bullets started flying my way.

*"you are with **them**, aren't you?"*

The war is between the two largest communities in theoretical physics, the *string theory* and the *loop quantum gravity* communities. To say that they don't get along is an understatement.

And Johannes and I didn't know. We placed ourselves in the middle of this war. We walked in with our jolly little shovels and colorful plastic buckets, our pink umbrella, and plastic chairs, and started building our sandcastle, hoping to find playmates. Right there between the two opposing armies, in no man's land, with no army of our own and with no military training whatsoever.

✡

Adam is almost finished. He is on his last slide, just a conclusion, and then I'll stand up and ask if anyone in the audience has a question for him.

I've invited Adam to give a talk at the Niels Bohr Institute. We're in the famous auditorium A. The room is nearly empty — perhaps twelve or fifteen people — and Adam has almost finished his talk on noncommutative geometry and the reconstruction theorem. He believes he has a proof of the equivalence of Riemannian geometry and Connes' spectral triple formalism and I got the idea that it would be interesting for the theoretical physicists at the Niels Bohr Institute to hear this. Something different, I thought, fresh blood perhaps, news from the moon.

And something else. I've noticed that there is very little contact between the institute of mathematics and the Niels Bohr Institute. I know that some of the mathematicians have a real interest in physics, so I thought that I could provide a link, make myself useful. Instigate a connection of sorts.

I've asked Adam to mention Connes' work on the standard model. It's still very new for me. I'm trying to read Connes' papers, but they are tough reads. Adam is a mathematician, his field isn't the standard model, but he said yes, "*of course, I can do that, no problem*".

It is less than a year since I arrived here after my two years in Iceland. That must be the reason for my mistake, I haven't learned it yet. To be careful, to know that you should *never* talk about something that you're not an expert on, never expose yourself. Keep up your guard.

After Adam's talk, there are several questions about the mathematics, and then there is a technical question about Connes' work on the standard model. I am the one who invited Adam to give a seminar, so I have the role of moderator. Adam doesn't know the answer to this question, I don't know it either, I'm no expert, I just think it all sounds incredibly interesting. I'm so naive.

The audience won't let it go. They start to discuss among themselves. It's a problem to tell physicists, who are experts on particle physics, something about the standard model and then not be able to answer their questions. It's a mistake and it's my fault. I should never have asked Adam to talk about physics. The audience keeps going. They want an answer, but Adam doesn't have it and eventually I ask them to leave the question unanswered. To let Adam off the hook, he is no physicist after all.

But I am.

Afterward, I meet one of these inquisitive professors in the corridor outside his office — two doors down from my office. He immediately starts telling me how outrageous it was that Adam didn't know the answer. He is really upset. I can see the anger in his eyes. I'm shocked. How can he be so upset about a question? I try to defend Adam:

"yes, I understand, but it was my fault, you know, I was the one who asked him to mention Connes' work, so if anyone is to blame it should be me", I tell him.

He then tells me that the answer to the question is obvious, and he keeps going on about how incompetent mathematicians are

"... and don't bother me with more of this stuff of yours, I'm not interested, and don't organize any more talks like this one", he tells me and walks away.

I am in shock. He's one of the most respected theoretical physicists in Europe, I've always admired him for his deep knowledge of quantum field theory. I'm standing alone in the corridor not knowing what to do with myself. Why had he insisted that Adam should answer the question when he knew the answer already? And why did he become so angry? In his eyes, I had seen something that looked like contempt. My hands are shaking.

I don't show up for work for the next few days. I feel awful.

I want to quit. I want to let it all go, leave this charade and stop pretending that this is ever going to work, that this idea will ever fly. Who am I fooling? It will never happen.

The thought of walking away from it all seems so incredibly tempting. But I don't quit. I can't. Not as long as I still believe in it.

The Ardez gorge, Austria, some years earlier

I can see Christian on the other riverbank. He waves at me and signals which line I should take: across to the large boulder in the middle of the river and then down towards him, just before the waterfall that he has been scouting.

I get back into my kayak, close the spray skirt, and grab my paddle. I sit there for a second or two before I put on my nose-clip and shift my balance with my hip so that I skid down the rocks into the water. I lean to my left and with a series of quick strokes enter the main current and maneuver across. As I get to the calmer water behind the boulder I pause for a moment and look across my shoulder to read the current downstream. With two strong strokes I then break through the wall of water in front of me and lean heavily downstream. I reach high and place my paddle above my head vertically into the strong current and use it as a lever to pivot myself around so that I end up facing downstream. I immediately pick up a lot of speed and it takes some effort to reach Christian just before the waterfall.

He gives me a thumbs up and signals to me to get out of my kayak and come with him to scout the waterfall ahead. The roar of the river is intense here, we only talk when close.

We're in the middle of the Ardez gorge, the most difficult whitewater section of the Inn river in Tyrol, Austria. There are three of us: Christian, Michael and me. Christian is the only one of us who knows this part of the river. He is the oldest and most experienced. I know he can be reckless, but I trust him.

We started this morning with the Giarsun section, which ends where the Ardez gorge begins. Vera, Michael's girlfriend, had been with us then and it had been a swift run. According to the book, the Giarsun is a grade IV river

---

170

but I had been surprised by its power. Several times I had struggled to prevent my kayak, which has a flat back section constructed for performing tricks, from backflipping in the strong currents, and Vera, who's pushing the limit of her technical skills in a grade IV river, had taken an ugly swim in the icy cold waters.

The Ardez gorge is listed as a grade V-VI river, which means that some parts of it are at the limit of what can be paddled. There is real danger here, if you make a mistake in the wrong place it could be fatal.

But Christian had told me that it would be fine. He knows me well and had convinced me. I wanted it too, of course, I did, whitewater kayaking is like a drug, and once you're addicted you constantly want to increase your dose.

I get out of my kayak, secure it, and climb a few rocks to follow Christian down the right riverbank to take a look at the waterfall. It's not that big, perhaps three or four meters, quite wide and regular, but the back current beneath the drop looks ugly. If you get trapped there, I think you might not be able to escape.

Michael stands next to his kayak a few meters further down the river. He has already decided to skip this drop. I'm also reluctant as I discuss the line with Christian, who is smoking a cigarette.

In the end, Christian decides to run it while I join Michael. We get our throw-bags with ropes out and try to position ourselves to best help Christian if he gets in trouble. And then we wait for him to go over the edge.

His kayak is very small and when he goes down the waterfall he completely disappears for a few seconds.

We're ready, but the truth is that if he gets pulled back into the fall he's going to be on his own — we can't reach him from where we stand. But Christian is fine, when he reemerges in the middle of the back current, he easily manages to push himself through.

This is the most intense river I have ever kayaked. The level of concentration is very high with little room for mistakes. I am always scared when I get into my kayak to paddle a difficult river — my hands shake, my stomach rebels, my mouth is dry — but once I'm in it, it's as if my body knows that fear will not help anymore and instead it offers acute alertness beyond anything I have experienced elsewhere. It's the kayaker's high, and right now I'm higher than I have ever been before.

As we continue working our way down the river, we reach a series of waterfalls followed by a few hundred meters of very technical whitewater. Christian goes first and I follow his line. But as I go over the first drop, I miss the line and a sudden side current flips my kayak to the left. I fail to react quickly enough and immediately find myself upside down going over the edge of the waterfall. A rock scrapes my helmet as I quickly lean forward, twist my paddle around and make a sweeping stroke that flips me back up again just as my kayak plunges into a small pool before the second and larger drop. The whole thing takes little more than a second and I immediately begin to paddle frantically to get momentum for the next drop.

The power of the current is amazing! I knew that this river would be difficult, but I didn't expect the force of it. In the book, it is described as a very technical and partly dangerous river but what I experience now is different — it's like dancing with a dragon or surfing an avalanche, it's beautiful and intoxicating and crazy but right now I know that I'm surfing my own limitations too. This is ballet on a battlefield, it's my life's solo with grenades exploding around me. The present moment, the symphony of a perfect stroke, expands and expands till there is nothing left but this waterfall, these rocks, my hands gripping my paddle, my hips shifting weight, this wave that pounds my chest and the roaring abyss that opens to my left and threatens to swallow me and everything I believe in. It expands till there is nothing left but right now, this moment with my brothers in arms and our intense will to breathe.

According to the guidebook, this run should take five to six hours, but we're through in just three. At the end of the run, we meet Vera, who is waiting in a parking lot by the river with one of the cars. We get out, get changed, load the boats onto the roof of the car and then head for a local *Gasthaus* in the nearby village for a beer and some food.

And there, sitting around the table drinking the first beer, Christian tells us what he really knows about the river, that we have just gone down.

He's smoking and laughing. The full effect of the drug kicks in after the run.

"Guys, do you know how much water was in the river today?"

He looks at us, laughs again, and continues

"it was in flood! The water was so high, I think that there were almost two extra meters in there today! Perhaps more! That's why it was so fast, it must have been all the

rain they've had in Switzerland. The river was so high, my God, I hardly recognized anything at all!"

Michael and I are stunned. None of us speak, we just look at Christian in disbelief. When you kayak a river in flood, you're kayaking a completely different river. It can be dangerous and usually requires scouting and preparation. And to kayak the Ardez gorge with so much water in it — without knowing it — is reckless at best.

"But why didn't you tell us?" I finally ask Christian and continue

"This was fucking dangerous; we didn't know what we were getting into — you should have told us!"

Christian looks at me with a smile, takes a draw on his cigarette, and laughs

"yes Jesper, I know, but if I had told you guys that the river was in flood, you'd never have gone with me. You would have chickened out!"

It is true. If I had known how difficult it would be, how dangerous it was, then I would never have done it.

But I am glad I did.

CHAPTER 17.

THE LOOPY STUFF

The stars went out
one after another,
when they found
the so-called philosophers stone

— The Legend of Josha and Ming
Gasolin, 1974

How do you quantize a theory that cannot be quantized?
Like gravity, for instance?

Since the standard quantization procedures do not work
for gravity and since we have no direct experimental data
to guide us, there is an empty canvas for ideas and
approaches. Whatever you can think of that satisfies a
number of the key aspects of a quantum theory — but not
necessarily all of them — can be regarded as a possibility
worth exploring. It's like you're given a half-finished
painting to complete: you know what the lower section
looks like, the ground, a few bushes, and a rabbit — but
what the rest should look like is entirely up to you.
Except, of course, that one day your painting will —
hopefully! — be compared to a landscape that exists out
there somewhere.

This empty canvas situation has led to the pursuit of several extremely different ideas over the past decades. By far the most popular has been string theory, which proposes that everything consists of tiny vibrating strings that live in 10 dimensions and feed on an exotic mathematical substance known as supersymmetry. But there are other approaches too, some close to the standard quantization procedures and some quite far off.

I would like to discuss one particular approach that has attracted a good deal of attention. An approach, which in some ways lies close to the standard quantization procedures but nevertheless departs from it. This approach is not without controversy, but nothing in this business is, we'd better get used to that. We're not in the social domain of polite Victorian-style tea party conversations, so let's not worry about rumors and reputation. What I would like to discuss now is known as *loop quantum gravity*.

✡

When you apply the ordinary quantization procedure to general relativity your first task is to decide which mathematical formulation you will use as your point of departure.

In 1986 an Indian theoretical physicist named Abhay Ashtekar, who was a professor in physics at the Syracuse University in the United States, discovered a new formulation of general relativity, which renders it in a form very close to that of a gauge theory. We know that all the other fundamental forces of Nature are described in terms of gauge fields and with Ashtekar's discovery the last one, gravity, can be formulated in that same general formalism.

This does not imply that Ashtekar's formulation allows one to quantize gravity in the same way as the other gauge theories — a new formulation never solves problems just like that, it merely shuffles them around — but his discovery opens a door to a different approach.

A gauge field tells us something about how a 'stick' is moved from one point in a space to a nearby point. If we continue the process of moving a stick between nearby points along a path, we will eventually have moved the stick between points that are not neighbors. Such a process is called a *holonomy*. A holonomy simply tells you how you move a stick between arbitrary points in a space. And remember, this stick can live in an abstract space too, and in particular it can be the internal spin of, say, an electron (an electron is what is called a fermion, which always comes with a particular feature known as spin).

What Ashtekar's formulation of general relativity made possible was an approach based on holonomies as the basic variables. Thus, the new idea in loop quantum gravity is to apply the standard quantization procedure to variables, which are not fields as we are used to, but instead holonomies along *paths*.

There is something very appealing about this. These holonomies simply tell you how you move a particle like an electron from one point to another. This is a very basic physical operation, and it makes sense that such a quantity should play a key role in a fundamental theory.

And there is an important spinoff: a quantization based on holonomies does not require a background geometry the way the ordinary quantization procedures do. By using holonomies we have a possible way around the fundamental problem of background independence.

So, this is loop quantum gravity: a quantum theory where the basic variables — the basic *operators* — are holonomies.

However, this emphasis on holonomies entails several serious challenges, the most important of which is the task of obtaining general relativity in a classical limit. I.e., to *get back* to the starting point, which is gravity as we know it, formulated in terms of curvature of ordinary space and time. To find such a limit in the loop quantum gravity approach has proven to be a seriously difficult problem.

This problem can be traced back to the way the Hilbert space is constructed in loop quantum gravity. Recall that a quantum theory is a combination of two key elements: an algebra of *operators* and a Hilbert space, where the operators can *act*. Without the Hilbert space, you don't have a theory. In loop quantum gravity you start with an algebra, where the holonomies play the main part, and you need to find a Hilbert space for this algebra. What Ashtekar and his coworkers did was to build one by taking a certain limit of *graphs*. This means that they consider holonomies along the edges in finite graphs (a graph is a collection of points connected by a number of lines or paths) and the limit where the number of *all* possible graphs becomes infinite.

In terms of the mathematics, this is completely solid. It can be done, there is no question about that. But the Hilbert space that you obtain from this procedure has a built-in structure that emphasizes isolated *points* in space. An ordinary space is, in mathematical terms, not simply a collection of *points* but carries also information about which points are *neighbors* — this is what is called topology[4]. But this information is lost in the process that is used to construct the Hilbert space in loop quantum gravity. The problem with this loss of information is seen when you look at general relativity, i.e., the classical theory, where the information about neighborhood is essential for the theory to make any sense. So, to get general relativity back the information must somehow emerge in the classical limit of loop quantum gravity, and this is precisely where the difficulties arise. The information about neighborhood, once lost, is very hard to reinstall in a quantum theory. The problem is, essentially, that space becomes *discretized* in the setup that loop quantum gravity uses.

On the other hand, this is precisely what the researchers behind loop quantum gravity celebrate about their approach. That space truly looks like a quantum reality in their formulation and that quantities such as area and volume come in discrete quanta much like the electron orbits in the hydrogen atom. What they find is a *granular* space.

And perhaps they are right, perhaps a theory of quantum gravity truly looks this way, but we need to remember three facts:

1.   we don't *know* anything about what a theory of quantum gravity *should* look like. It may have a granular structure of space (and time?) or it may not,

---

[4] To be precise, what is lost in the construction of the Hilbert space in loop quantum gravity is information about the *differential structure* of the three-dimensional manifold. The differential structure enables you to form differential operators, i.e., rate-of-change operators, which are a key ingredient in any field theory.

we simply don't have *any* a priori knowledge of what such a theory ought to look like;

2.   the discretization of space that occurs on loop quantum gravity is essentially the product of a *mathematical choice* made when the Hilbert space is constructed — i.e., it is essentially an input, not an output — and

3.   the need to find back to the classical theory of gravity in a classical limit is absolutely critical.

- But if you believe that the construction of the Hilbert space in loop quantum gravity is problematic why not just do it differently?

Well, that is the thing, it is not so obvious how else to proceed. The Hilbert space matches, in a certain sense, the algebra that it hosts. And if you start to question the algebra based on holonomies then you're questioning the very basis of the loop quantum gravity approach. The holonomies. If you change that you change everything.

This is precisely what Johannes and me are doing in our work. We'll get to that shortly, we just need to cover a little more ground first.

✡

Johannes and I visited Abhay Ashtekar at Penn State University, where he is now a senior professor, in the fall of 2009 to discuss our work and the correct implementation and interpretation of Ashtekar variables with him — these variables are now named after him.

Ashtekar is a very friendly man, he is not so tall and has a round face with a genuine smile that somehow reminds me of my travels in India, where I have always admired and appreciated the warmth of the people.

Johannes and I had visited the Perimeter Institute in Waterloo just west of Toronto and had hired a car to drive down to Penn State. It was mid-October, and the scenery was spectacular. I spent most of the drive talking about the beauty of the fall colors, how incredible all these maple trees looked as they surrendered themselves to the almanac, until Johannes finally told me that he is color blind and that it all looked brown to him.

We spent a few days at Penn State where we had discussions with Ashtekar in his office. People in the loop quantum gravity community can — in my opinion — sometimes come across as being somewhat unwilling to discuss the basic setup of their theory; I have at times sensed a defensiveness, as if they fear to expose weakness, but Ashtekar proved to be an exception. He was very open-minded and told us about his considerations on how to quantize gravity using his variables. He readily admitted the weaknesses of his approach and said that he simply had no other ideas. It was clear to us that he had a deep understanding of the mathematical issues at hand.

Ego often plays a large role in theoretical physics — I suspect it does in most of science — and often I find it to be a hindrance: the truth is that we all seek the same thing and that we need all the different takes and angles on these problems as we can get. When egos get in the way we tend to shut down our creativity and our sense of comradeship and we stop listening. I don't know if it is his Indian roots, but Ashtekar came across as a man who was not burdened by his ego; there was something youthful and joyful about him.

This openness and joyfulness is something I have often noticed when talking to the most gifted scientists, and this in particular when it comes to mathematicians: they seem to simply enjoy their play with structure as if their intellect is most happy when it has freed itself completely from physical bounds and been let loose in the realm of pure reason and logic.

✡

According to what I have written so far, the basic ingredient in loop quantum gravity is the holonomy, which tells you how to move a particle with a spin (could be an electron) from one point in space to another.

But this is not completely accurate.

Loop quantum gravity *could* be based on holonomies had it also involved matter, but that is not how it is set up. A priori it is a theory of quantum gravity *only*.

Instead, what loop quantum gravity is based on is a certain *contraction* (the mathematical term is 'trace') of holonomies, which extracts the information that is relevant when there is no matter. The holonomy itself is a matrix — i.e., an array of numbers — and its contraction is a *single* number.

But here is the thing. If you take an algebra based on holonomies, then it is noncommutative. It matters, critically, in which order you multiply them. But if you use contracted holonomies, as is done in loop quantum gravity, then the algebra becomes *commutative*. It's cut down to — essentially — just numbers.

This step, the move from holonomies to their contractions is what caught my attention when I began to read loop quantum gravity review papers in my office in Reykjavik in 2003. It was one of those early autumn days with mixed weather, where you see a sunny afternoon when you look out the window but find a snowstorm when you walk out the door. I was flicking through a paper written by Ashtekar and Lewandowski when I saw it.

I read about the contraction of the holonomies and that made my heart miss a beat.

Because that was exactly what I was looking for. I recognized it immediately. This was it, I thought, it was completely obvious.

I was looking for a possible mechanism that could help explain the mathematical structure that Chamseddine and Connes have identified in the standard model, the *almost-commutative algebra*, and I had come to the conviction that one should search for such a mechanism in a framework of pure quantum gravity. And that day in Reykjavik I believed that I had found it.

The point is that if you work with the holonomies themselves (and not their contractions) then you're automatically in the realm of Connes' noncommutative geometry. The algebra of holonomies is *noncommutative* and will therefore automatically activate some of the mathematical mechanisms that push you towards unification, i.e., a theory that does *not* — that *cannot* — only involve gravity, just as Chamseddine and Connes had found.

I thought it was so obvious. The loop quantum gravity approach carries within its very core an *unrecognized key ingredient of unification*, a dormant seed with important mathematical information, which had so far just been thrown away.

And how beautiful that would be if the mathematical structure of the standard model itself would emerge from a theory of *pure* quantum gravity. If the machinery that Ashtekar and his coworkers had found could eventually provide the answer to *it all*. The standard model *and* quantum gravity. It would be such an elegant solution. If Nature had chosen this path, I would think she had chosen well. I immediately fell in love with the idea.

I was so excited, I got my jacket and went out into the snowstorm, where I headed for the Nordic culture house for a coffee.

Of course, things are never as obvious when you start to look at them in detail. That is why I always leave my work when I get a good idea. I want to *enjoy it* a little while, the *feeling* of the idea, the joy, the relief, before I start to pick it apart and find all the problems.

And with this idea problems did indeed emerge. Lots of problems. It took us more than ten years until we began to see how it could be realized.

CHAPTER 18.

AFRICAN HITCHHIKING

"I have found out one thing and that is,
if you have an idea, and it is a good idea,
if you only stick to it you will come out all right."

— Cecil Rhodes

Malawi, Africa, April 2010

I hear it before I see it.

A deep, rhythmic sound that anyone who has seen Vietnam war movies will recognize.

I hear the sound first. My brain freezes. It makes no sense. Out here? How can that be? *A helicopter*? Impossible. Or? Could it be? No! No, no, no, no, please no, it can't be, it mustn't be, please God!

The helicopter emerges at the mouth of the bay. It flies about 200 feet above the surface of the lake and is heading straight north. It is huge, army-green, and with a single rotor. I pray it will continue north but I know that it won't. My brain is no longer frozen, it is on fire, all my neurons are popping. I know that the military helicopter out there over Lake Malawi has come for me.

When it reaches the heart of the bay, the helicopter makes a smooth, wide turn and heads straight towards me.

✡

Four days earlier

My brother Simon is standing at the top of the cliff. I can see from the way he stands there that he has lost his nerve. He is looking nervously down the route he climbed up. It is an easy ten-meter climb, but I know that my brother is afraid of heights and I know that he now regrets his urge to show off.

I swim out to the little island, which is about thirty meters off the shore of Lake Malawi and start to climb up to where he stands. I can see why he is nervous — to jump off the cliff you'll have to clear a couple of rocks protruding from the water. It is an easy jump but if you fail it will be the end of you.

To help him I decide to jump first and show him that it is not difficult. I leap off the cliff and hit the water a couple of seconds later. I am surprised by the force of the impact. It really hurts, especially in my neck. The pain runs up my spine.

I swim back to the island and look up at my brother. He is hesitating, he isn't going to jump.

My neck hurts but I climb back up to my brother and together we manage to climb down five meters to a place where it is easier to jump. My brother goes first, I jump second.

When I hit the surface of the lake for the second time my neck makes a strange sound. It hurts a lot and when I surface, I have difficulty swimming — I lay back in the water and float. Luckily my brother notices and he manages to get me back on shore. The pain in my neck is intense and I have trouble moving.

✡

We have hiked for four days along the shores of Lake Malawi in Africa. My brother and his friend and me. We had been bored at the backpacker's place, where we spent a few days, and had decided to just head north. *"Let's see what's up there"*. We passed many villages on our way and camped out on beaches. In the beginning, we had walked close to the water and crossed the muddy rivers on foot but slowly we realized just how much the locals feared the crocodiles and we began to keep our distance.

Finally, we reached this little village in the middle of nowhere and its small, empty hostel at the shore of Lake Malawi.

I had finished my job at the Niels Bohr Institute a month earlier. My application at the Carlsberg Foundation had only been partially successful: they granted me one year's of salary instead of the three years I had asked for. None of my other applications had been successful. Not even an interview anywhere. Nothing. It had become increasingly clear to me that my career was coming to an end. To postpone the inevitable I had moved the starting date of the Carlsberg grant a few months ahead and flew to

Malawi to meet my brother, who has a teaching job in Mozambique.

And now I am lying here on my back with three books stacked under my head to ease the pain. I can hardly move, and I am worried. Since my childhood, I've had plenty of problems with my back but nothing as bad as this. It feels serious.

The pain has not eased after a night's sleep and the next morning we decide that we have to do something. I am pretty sure that nothing is broken — I can move my feet and feel no unusual sensations in my lower body — My guess is that I have some kind of disc displacement somewhere in my neck. There are at least four days of hiking to the nearest road, but we know that there is some sort of radio in a village about a day's hike south of where we are. My brother, who is a strong long-distance runner, sets off. The idea is that he'll try to get hold of some strong painkillers and get a signal out to my insurance company by way of the radio so that they can help organize a doctor with some morphine on a boat.

This turns out to be a really bad idea.

When my brother returns that evening, I feel a little better and can now stand up. He tells me that he tried to use the radio but isn't sure if it had worked. Since I still can't walk, we decide that my brother and his friend will continue north the next day and try to organize a boat so that they can pick me up later. I'm supposed to stay at the hostel and wait for them.

And so I do.

✡

It seems like ages ago but only two months have passed since we had the Oberwolfach workshop in February, when we brought together experts from noncommutative geometry and loop quantum gravity to discuss our work and the idea of an intersection of their two research fields. It was in the middle of winter; the snow was packed two meters deep.

It had been my idea, the workshop. You only go to Oberwolfach by invitation and the workshops that take place there are selected by a committee based on applications. Johannes and I thought that it wouldn't be proper to send an application in our own names, when the workshop that we wished to organize was to be on our work, so we had instead written an application with the German mathematician Christian Fleischhack and put him down as the official organizer along with two others. It was positively reviewed, and we got the workshop.

25 people, one half from loop quantum gravity, one half from noncommutative geometry, one week. It had been intense, sometimes the debates were heated, even hostile. Bringing people from these two different fields of research together turned out to be somewhat explosive, each community has its social structures, its hierarchies, its kings and queens and peasants too — I saw it a bit like two hedgehogs mating: painful and difficult.

The atmosphere at the workshop took a dive when one of the leading physicists from loop quantum gravity gave a talk. It was the young German professor, whom Johannes and I had visited in Potsdam a few years before, who was to introduce their research field. A famous mathematician from the noncommutative geometry camp sat in the audience and halfway through the talk he launched an attack. He demanded some precise answers and he wanted them now. The scene had developed into an outright interrogation, where the mathematician stood at the blackboard questioning the German physicist while the audience watched them. I was impressed with the German's calmness. I do not know if he actually has military training, but I have never seen anyone withstand such an attack with his level of calm. I think that this provoked the mathematician even more and after that, there was a state of war between the two.

One day when the whole group had visited a nearby *Gasthaus,* the two started fighting again. As the unofficial organizer, I felt some responsibility, especially when the mathematician started shouting at the German, but I knew that any attempt to mediate was doomed to fail.

✡

On the fourth day of the meeting, I took a walk with one of the leaders of the loop quantum gravity community, a Polish theoretical physicist. I liked him, a very straight forward "what-you-see-is-what-you-get" kind of guy. He was very excited about our ideas, the possibility of applying the mathematics of noncommutative geometry to a setup similar to what they use in loop quantum gravity. I got the impression that he had been waiting for something like this to happen, that he was perhaps more inclined towards pure mathematics than most others in his community and that he now, finally, saw an opening towards a *cleaner* approach.

During our walk, I told him about my job situation, that I didn't have a secure position and that I was applying everywhere and mostly in loop quantum gravity strongholds. That seemed natural, no one in string theory will hire someone like me, so a loop quantum gravity position seemed like my best chance. But when the Polish physicist heard this, he abruptly told me that I had zero chance of getting a job in their camp.

- *"You'll never get a job with us, it's not going to happen, we only give jobs to our own,"* he said.

I was amazed by his candor. His remark was not even hostile, just completely honest. I had never been told something like that, the inner workings of the different research communities are somewhat enigmatic. But here it was, the truth: I'm not one of them and therefore I'll never get a job in their camp.

My heart sank. At that moment I realized that I would not be able to stay in physics as a paid researcher. It was so obvious: theoretical physics consists of tribes and each tribe has its own hierarchy. You make a career *within* a tribe by working your way up. And if you don't belong to one of the tribes, you're just not going to make it. You're an enemy, or a nobody at best, an anomaly. I had been blind to this; I didn't see it before this Polish physicist told me.

After our walk, he invited me to a large loop quantum gravity meeting in Krakow in Poland a few weeks after our workshop. It was the annual meeting where all the loop people from around the world meet to discuss their work. I had accepted the invitation on the condition that I would be able to present our work there. At first, he had refused — apparently only insiders get to speak there — but later on he accepted.

In the end, both Johannes and I had gone to Poland where I gave a presentation. I don't know what difference it made, though. There had been some discussion after my talk but then the young German professor, who had also been at the Oberwolfach meeting, stood up and proclaimed that our approach was *obviously wrong*. After that nobody said anything.

✡

My neck hurts. I carefully raise my arm and reach up to adjust the books. I'm doing better, I can walk around a little, but most of the time I'm lying down with a stack of books under my head.

My mind begins to wander off again and I recall the time back in Iceland when I first got the idea and how excited I had been, how I had expected the whole world to jump on it. I had told Johannes that we had to be quick, that once this idea was out others would pick it up and outcompete us. That we would only have one shot at this. How wrong I had been, there is no competition here, we've got this whole damn sandbox completely to ourselves.

✡

On the last day in Krakow, Abhay Ashtekar had given a talk. Johannes and I sat in the back of the auditorium and were curious to hear what he would talk about. I hoped that he would discuss his views on the classical limit of loop quantum gravity, but instead he spent his time talking about strategy – how their community could best position itself against the string theory community, and how important it was that everyone should call their research loop quantum gravity instead of other things. He admonished everyone to stay united and to appear undivided.

Afterward, Johannes and I talked about how closed the different tribes in theoretical physics are. He pointed out the stark contrast to mathematics, where people do not cluster the way they do in physics. It reminded me of the knight Galahad in the legend of the holy grail, for whom it had been a question of honor to enter the forest in a place where it was at its densest; to follow in someone else's tracks was for him unthinkable. Physicists don't think like that; they seem to prefer the highway. Why crawl around in the mud in the jungle when you can sit comfortably in your car and listen to the radio while driving alongside everybody else?

I'm lying on my back watching the dense jungle through the open window. I can hear faint noises from the village and loud screams from some kind of bird.

Something I experienced at the latest Christmas dinner at the Niels Bohr Institute comes to mind. It was towards the end of the evening when I sat beside a Ph.D. student, who had asked me what I do. I gave him a brief outline of what I am working on and when he seemed interested, I gave him more details.

At one point he interrupted me. I remember his exact words, said almost laughing: *"but are you allowed to do that?"* He was surprised that I am working on my own project, my own ideas. The thought that you can do that apparently surprised him.

Afterward, I had thought about this as a funny anecdote that speaks to the fact that students nowadays learn a lot of techniques but are not taught that doing research also involves being creative. The point is not only to dig a hole as deep as possible but also to find the right spot to dig your hole. But now I can't help thinking that he had been right. The answer to his question is that *"no, you are not allowed to work on your own ideas"*.

I'm staring up at the bamboo ceiling above me. I can see a big spider just above my head. I am thinking about the story of Galahad. I don't know if he asked anyone for permission when he mounted his horse to ride off in search of the grail. I don't think so. I don't think he cared.

The helicopter crew sits in silence around the table and stares at me. There are four of them, two pilots, a mechanic, and a doctor, all Malawi nationals. I tell them that I no longer need assistance and that I prefer to stay here " ... *but thank you very much for coming* ... " I say and try to smile. I can see that that doesn't go down well. One of the pilots insists that they have instructions to take me with them back to the capital. And from there further on to Johannesburg in South Africa.

When the helicopter landed on the beach by the village my first thought had been to hide. Or run away into the jungle. I knew what had happened: my brother's radio signal had reached my insurance company who had hit the red alert button. The plan had been something with a boat and a local doctor, but somehow that turned into a huge military helicopter and a complicated international transfer to a South African hospital. And all I can think of is the bill. Who will pay for all this?

The helicopter crew has flown for two days to reach me and I can tell from the looks on their faces that there is no way they will leave this place without me. So, I gather my stuff and go with them. As I walk along the beach towards the helicopter carrying my backpack the entire village watches, perhaps 500 people.

I know that many of those villagers have AIDS and I know that they have never seen a helicopter before. There is a total of five helicopters in Malawi, all military, and none of them will ever come to this part of the country unless someone from the outside is paying for it.

This is the most embarrassing moment of my life. To walk along that beach in front of an entire village and climb into that helicopter. I am embarrassed for the mess we have made, and I am embarrassed for the sudden display of just how much more valuable my life is compared to the lives of these locals: I have a sore neck and a freaking AS532 Cougar transport helicopter with a medical crew drops out of the sky to pick me up. These people are dying of AIDS and nobody will send even a postcard.

Welcome to the real world. This puts my career problems into perspective.

✡

10 days later

The noise is incredible. The sound of the rain pounding on the tin roof sounds like machine guns; it sounds like nothing I have heard before.

Simon, my brother, is lying on the floor, he is covered in sweat. I can see that he is shaking too. He looks very sick.

I have just been down to check the river. The torrential rain drenched me in seconds. The river is very high, there is no way we can cross it today. Perhaps tomorrow, if the rain stops, perhaps, but I am not certain.

We came up here yesterday, in the Mulanje Mountains in southern Malawi. We are sitting in a hut a few hundred meters below the peak. The hike up here was easy, it took us about a day where we had to cross a couple of small rivers on the way, nothing serious. Simon had been okay then, he told us that he felt sick and thought that he might have malaria. He had taken a large dose of Malarone and said that he would be fine.

At the hospital in Lilongwe, the capital of Malawi, they had given me some very strong painkillers for my neck, and yesterday before we started the hike up the mountain, I gave Simon a handful of them. Just to ease the pain. It was a really stupid thing to do, you don't hike up in the Mulanje Mountains if you feel sick.

And you don't take painkillers if you think you've got Malaria.

And now the Australian man who shares this makeshift hut with us for the night has just told us that Simon could have meningitis.

- *"He has the symptoms,"* he said

- *"it could be meningitis."*

His fever is very high. I don't have a thermometer, but I don't need a number to see that he is very sick.

It's odd, but there is a phone connection up here. When I switch on my phone, I see four green bars indicating a solid signal. So far up and yet we can call home, or anywhere we want.

I am sitting outside the hut with my back against its wall. The roof is wide, the downpour does not reach me. I've got my phone in my hand.

Shall I make the call?

I don't know much about meningitis, but I know that it kills you. Quickly, you don't have much time. But does he have it? How can you tell?

But I can't call in yet another helicopter! I simply can't. What if it's just the flu or one of the thousands of other illnesses you can get in Africa? It would be so embarrassing.

I sit there a long while looking into the rain.

Then I turn off my phone and go back into the hut to my brother.

CHAPTER 19.

FROM CLASSICAL TO QUANTUM AND BACK

"Some people say, "How can you live without knowing?"
I do not know what they mean.
I always live without knowing. That is easy.
How you get to know is what I want to know."

— Richard P. Feynman

What is primary, the classical or the quantum?

Until now we have discussed quantum mechanics and quantum field theory in terms of a procedure known as quantization. In this procedure, classical quantities such as the position and momentum of a particle are replaced with operators and thereby elevated to a machinery of algebras and Hilbert spaces, i.e., a quantum theory.

This procedure starts in the *classical*. The quantization procedure is applied to a *classical* theory, for instance, the electromagnetic field or a system of particles with some interaction, and the quantum theory emerges as a function of this *classical* setup.

Once you have a quantum theory you then need to check that you have a *classical limit* where the theory that you started out with reemerges. This is necessary because the world that we experience around us is classical and must be included as a limiting case.

There are also important consistency checks that one must perform in a quantum theory, which tell us whether the symmetries that apply to the classical theory also apply to the quantum theory. Some of these checks are critical; if they fail the theory is considered meaningless.

This means that there is a set of operational arrows that take us from **classical** to **quantum** and *back* to the **classical**. This pattern applies to essentially all the quantum theories that describe Nature. This means that these quantum theories are motivated by their classical counterparts. The choice of fields, the choice of interactions, the primary *conceptional* input is essentially classical.

This is the case for loop quantum gravity too, where general relativity, i.e., the *classical* theory, is (attempted to be) quantized. If you view loop quantum gravity from the perspective of the quantum theory *alone* — i.e., the choice of algebra, the way key operators are constructed and the choice of Hilbert space — then it has a highly contrived form, which you would *never* come up with if you didn't know the classical theory beforehand. The mathematical structure of loop quantum gravity simply makes very little sense when viewed solely from the quantum side of the aisle. It only makes sense because it matches the classical theory at a structural level. Its justification lies in the *classical*, which in this case is therefore primary.

This is also the case for Chamseddine and Connes' formulation of the standard model of particle physics —

---

in a sense even more so. Here the process of quantization is only applied in a secondary step, after the standard model has been derived from the spectral triple construction, and it is only applied to one half of the construction: the gravity part is left out (which is understandable since we don't know how to quantize it). So again, the classical is *primary*.

And take ordinary quantum mechanics for that matter. You get quantum mechanics by quantizing classical mechanics: all the key operators are obtained by interchanging classical quantities with the corresponding operators. The classical is primary, the quantum is secondary.

But here is the question: what is more fundamental, the classical or the quantum realm? Clearly, the answer must be the quantum. The physical reality is primarily quantum mechanical and the classical world, which we experience, is an emergent effect — it is a limiting case. That *must* be how quantum mechanics should be understood[5].

So, if we are searching for a fundamental theory and if we believe that this is a quantum theory, then should it not be something that makes sense *in its own right*? Something that is not motivated or justified by a classical theory, but which is primarily a quantum theory?

The point is that ultimately the operational arrows must take their point of departure in the quantum world. We

---

[5] And of course, whenever there is something in quantum mechanics that seems certain, there is controversy, which is also the case here. The question of what is primary, the classical or the quantum, is a matter of interpretation and therefore clouded in the century-long debate on how to understand quantum mechanics. But I can say this much: *my* view is that the quantum must be primary.

must find something that *naturally* incorporates key features of a quantum theory, something that can stand on its own feet without being justified by a classical theory.

Think about Einstein's theory of relativity. This theory is not motivated and shaped merely by its ability to deliver Newtonian physics in a limit — no, its motivation, its *justification*, comes primarily from itself, its naturalness, its beauty, and yes, from its ability to deliver hard experimental and observational predictions, but that is not the point here. What matters is that Einstein's theory makes sense *by itself*. Newtonian physics is secondary. And the same must be the case for a final theory. Such a theory *must* make sense on its own terms. The classical world, the physics that we already know, *cannot* be its only or even its primary justification.

So, what we are looking for is *not* a quantization of general relativity or some other classical theory. That would be using the old operational arrows. What we are looking for is something that makes perfect sense *in itself* — both mathematically and conceptually — and something that naturally involves the key elements of a quantum theory, one of which is the ability to deliver the classical world in an appropriate limit.

We are looking for something that explains *why* the world is quantum mechanical. Why quantum theory? That is what we would truly like to understand.

We are beginning to see the program. It is time to jump off the cliff.

CHAPTER 20.

THE IDEA

*"Scientific progress is measured in units
of courage, not intelligence."*

– Paul Adrien Maurice Dirac

The trenches are far behind us. In the first chaos of the attack, we got separated from the main force, but we charge ahead regardless. The adrenalin is pumping through our veins, sweat runs into our eyes; we point our bayonet mounted rifles straight ahead as we run into the mist with a wild scream.

It is time to dig deep and look at the idea that Johannes and I have been working on for the past decade.

Let's start with a fundamental principle.

The most important ingredient in a quantum theory is the algebra of operators. This is the primary element and therefore the natural place to look for a fundamental principle.

Here is a question: what is the most elementary action you can think of when you are simply given a three-dimensional space? What comes to your mind? Consider this one: *moving an object from one place to another*.

Just that. Moving stuff around.

This is the idea: to construct an algebra from the basic mathematical operations of moving an arbitrary object between different regions of space and to build a theory around this algebra.

And note this: the objects themselves are not relevant — not yet — what matters is the *action* of moving stuff, whatever it might be. Not the stuff, just the *moving* of the stuff.

It turns out that an algebra built of such operations involves precisely the type of mathematical structure that we are searching for. Such an algebra will lead us to a theory, which is fundamentally quantum and at the same time involves a possible link to the standard model of particle physics. That is, a candidate for a unified theory of Nature, or if you like, a *final* theory.

So, let's take a better look at this algebra. To understand it we need to look at the mathematical process of moving an object from one region in space to another. Any kind of object. In particular, it may be an object with a form, with edges and spikes that point in different directions — really, anything you like.

To move an object with a shape requires information in the form of a *gauge field* — as we discussed in chapter 8 — and since we are moving it between two arbitrary points, we are talking about a holonomy (remember, a holonomy is the mathematical word for moving a 'stick' between arbitrary points according to the information given by a gauge field).

This setup is actually quite similar to loop quantum gravity, at least at the outset. The big difference is that we do not move single *points* around but instead objects, which have a finite size, which take up a finite *volume* in space.

Here is why: in all the theories that describe Nature (general relativity and the standard model, — all field theories), single points do not play any role *whatsoever*. What matters in these theories are finite volumes and finite volumes *only*. If you consider the electromagnetic field, it does not matter what the field does in a single point, it only matters what happens in a region of space (the mathematical term is that a point has zero *measure*). It is not difficult to understand why: there are clearly infinitely many points in space, so if it did matter what happened in each point and you were to sum it all up, it would always give an infinite result. The way to deal with this situation is to introduce a method of summing up all points, which is called measure theory, and when this method is applied to a single point it always gives zero.

It makes a huge difference if you consider the mathematical machinery that moves finite volumes around in space instead of points. Recall that one of the great challenges in loop quantum gravity is to recover the classical theory, general relativity, in a classical limit. The reason for this, as discussed in chapter 17, is that the mathematical information of *neighborhood* between points is lost in loop quantum gravity. Now, it turns out that this information is preserved when you work with volumes instead of points. This means that we will have access to a classical limit if we build our theory around this new algebra.

So, you can say that loop quantum gravity is also based on the idea of moving stuff around — i.e., holonomies — but what is moved around in that theory is precisely those elements that don't matter to physics, namely the parts that have zero measure.

Now, it is not enough to have an algebra made of elements that move objects around in space. There is a natural second ingredient, which changes the way these first elements move stuff around and it is the combination of these two ingredients, that form the algebra that we are interested in.

The name, that we have given this algebra, is the *Quantum Holonomy-diffeomorphism algebra*[6], or, in short, the QHD algebra.

Here are the two main reasons we are interested in this algebra:

1.   It includes the basic mathematical ingredient required for a theory of quantized gauge fields.

2.   It involves a possible link to the standard model of particle physics.

Let's jump into some details here to better understand these statements. Recall first our discussion in chapter 4, where we talked about the most important mathematical relation in a quantum theory called the *canonical*

---

[6] The mathematical term for the operation of moving stuff around in space is called a holonomy-diffeomorphism.

commutation relations. These are the relations that tell us how basic operators in the theory relate to each other.

It turns out that hidden within the mathematical structure of the QHD algebra one finds the canonical commutation relations for a quantized gauge field. This gauge field can either be one of the Ashtekar variables (which Abhay Ashtekar discovered in 1986) — in which case we are dealing with a potential theory of quantum gravity, or it can be an ordinary gauge field. This means that the QHD algebra has the potential to provide the platform for a theory of quantized gauge fields. This is, a priori, similar to loop quantum gravity but *not* identical.

And here is the second reason we find the QHD algebra interesting: when you formulate a classical limit and check what this algebra reduces to in that limit, then you find an *almost-commutative algebra.* That is, precisely the type of algebra that Chamseddine and Connes have identified as the central mathematical ingredient in the standard model of particle physics. The almost-commutative algebra is the key ingredient in the spectral triple, which they found to produce the standard model plus gravity — what we discussed in chapter 15.

Let's pause for a second to consider this statement. The QHD algebra has a priori nothing to do with the standard model at all, yet in a classical limit it produces the type of mathematics that Chamseddine and Connes have shown to be central to the standard model.

This means that there is a possibility that *everything* — the standard model of particle physics — could be the product of a theory that is only concerned with geometry, i.e., with space and time. There is a possibility that the standard model could simply be a *spin-off* from a theory that simply deals with how *stuff* is moved around in space.

This may sound crazy (and in a way, I think it is) but there is some logic to it. The QHD algebra is, primarily, related to gravity (how things are moved in space is a question of geometry), and Chamseddine and Connes demonstrate that the standard model *and* gravity together can be understood as a *single* theory of gravity. Thus, *both* the QHD algebra and Chamseddine's and Connes' work are about geometry and what Johannes and I now suggest is that they may be intimately related.

It is important to say here that the almost-commutative algebra that we find is not precisely identical — although it's pretty close — to the one Chamseddine and Connes find for the standard model. It is the same *type* of algebra. It has the same overall structure, but it is not *exactly* the same. There are good reasons why one would not expect them to be, even if the link between these two theories is real.

But the point here is the *possibility* of a link to the standard model. It is the *idea*. Whether this link can be substantiated and eventually verified is the subject of ongoing research and will — I believe — not be known for some time: it poses a delicate problem, and at the moment we have very little manpower to analyze it. But regardless of the outcome of this analysis, I believe that the mere possibility of a link, the *idea* itself, is worthy of discussion since it is one of the only existing candidates for a final principle. I am not aware of any other well-defined principle, which simultaneously has the characteristics of a final principle *and* generates a mathematical structure rich enough to make a connection to both the standard model and general relativity conceivable or even plausible.

So, let us for a brief moment forget all the technical details, all the accounting, bookkeeping, and nitpicking, and just swim in this idea: the possibility that *everything* — the fireplace that warms me as I write this, the cold wet winter outside my cabin window with the meadow and the forest and the entire universe — could be generated by a mathematical principle that simply encodes how stuff is moved around in space. This idea, this thought, continues to captivate me.

And here is why: the QHD algebra has precisely the key characteristic of a final principle that I advocated in the Shell Beach chapter, namely *emptiness*: the concept of moving an object is so elementary that it seems completely immune to further scientific reductions. The question *"what is the action of moving stuff around made of"* makes no sense. Of course, you can reduce it in mathematical terms — you always can — but at a *conceptual* level, it is a closed door. If Nature chose this algebra as the foundation of a fundamental theory, then I believe it must be the end of the road, that this is where the onion ends.

It will be the last turtle, if you like.

Note one thing about the QHD algebra. With it, the operational arrows between the quantum and classical realms that we discussed in the previous chapter take their point of departure solidly in the quantum. The structure of the QHD algebra is not dictated by a classical theory — or by anything else for that matter. There is no quantization here. In fact, one could argue that this algebra from the get-go has very little to do with quantum theory at all. It makes sense by itself as the algebra that encodes how stuff is moved in a 3-dimensional space. The quantum aspect is a consequence of this algebra, once you have decided to work with holonomy-diffeomorphisms (the moving stuff around), then the structures that entail the canonical commutation relations of a quantized gauge field are forced upon you. It cannot be any other way. This is precisely how we would like to see elements of quantum theory emerge. Not as an input but as an output, where the quantum is primary and the classical secondary.

Where quantum theory is an output; where we are given what might be read as an explanation of '*why quantum?*'

✡

— except, of course, that this explanation does not *really* explain it.

In the end, there is always a feeling that the *why*s we find don't *really* tell us why. This is the nature of questions, it is how explanations work. They always leave us with a trace of that childhood wonder — that *silent* wonder lurking in the shadows of our thoughts — as if what we are given is merely the scaffolding and that the building behind it remains forever hidden.

Forever out of reach.

✡

Johannes and I found the first trace of the QHD algebra in the fall of 2011 when I was a visiting researcher at the IHES institute in Bures-sur-Yvette just south of Paris. The IHES is a famous institute of mathematics that hosts no less than three Fields Medal winners, among them Alain Connes. The institute has a little park that looks beautiful in the fall and I spent much of my time there either walking in the park or sitting in my office looking out the window at the trees, all frozen in a magnificent explosion of autumn colors.

We were thinking about how to move beyond a construction that is based on lattices. We had spent several years analyzing different mathematical setups that all involved infinite systems of lattices and had become convinced that we were working with structures that had an existence independent of lattices and that we now had the information needed to uncover those structures.

Lattices are a computational tool — at least the way we viewed them — which means that it must be possible to avoid them, to peek beyond their dominion, and see what is going on, out there, in the realm of pure mathematics. It's like a scaffolding, you need it to build your house but at some point, you'll want to remove it so that you can see the building itself.

Sometimes it takes a very long time to see the obvious, it did in this case. But during one of my walks in the park at the IHES institute I got an idea about what we were dealing with and later that day I had a Skype conversation with Johannes where he immediately caught onto the idea and quickly worked out the details. What we were discussing was an algebra generated by the operation of moving stuff around in space — the *holonomy-diffeomorphisms* — it was very simple and we both felt that we had found what we had been looking for.

It took us another two years to find the second part of the algebra, the so-called conjugate variables. These variables, which correspond to the momentum operators in quantum mechanics, change the way the operation of moving stuff around in space works and we knew that they had to be included in our setup, but we did not understand precisely how. We felt that just adding them would be cheating. What we were looking for was a mathematical, logical reason why they *had* to be included. We wanted the structure to reveal itself to us instead of imposing our preconceptions upon it. We wanted the mathematics to speak to us instead of us speaking through it.

In June of 2013, I visited Johannes in Germany to discuss this problem. We spent a week working on it, but couldn't see the solution and then I left for the Dolomites, where I spent almost two months hiking through the Brenta mountains, a magnificent alpine massive in northern Italy.

During those weeks, when I walked from valley to valley carrying only a small backpack and an ice ax and stayed in mountain refuges and at alpine pastures, I kept working on the problem of the conjugate variables, tossing it around in my head, just as I had done in Tibet ten years before.

When I am faced with a problem like this I really try to think with my whole body. I try to *feel* the solution. I line up everything that I know in my mind and visualize it, like the landscape I walk through: the clouds, the rivers, the rocks and rolling hills. I try to smell it, sense it, touch it. I try to *be* it.

Because how do you find something that lies beyond what you know and what you can deduce intellectually? How is that even possible? How do you get a good idea? You cannot *think* your way across the gulf that separates what you know and what you cannot yet know. That is impossible, you must somehow imagine it, you need wings, you need to *fly*.

For me, there is a large element of trust in this, of leaning into the unknown, of trying to grasp something that I do not yet see. To allow myself to lose my footing and slip into an abyss of possible failure, misguidance, and a life wasted. It is the scariest and most exciting thing I know.

There is a trick that I have learned to use, a triviality perhaps, but important nevertheless: it is to believe, to *know*, that the solution to the problem exists, to count it *as a fact*. A fact just as important and useful as all the other facts that I have. And to believe that *I* can find it. The solution. That it is possible.

This is the extra piece of information I need to dare the jump, and to keep jumping, day after day, week upon week. To *know* the unknowable: that the solution is waiting for me out there somewhere and that it will, eventually, provide me with the grip I need to recover my balance. That it will save me.

It is a kind of professional faith, a rational irrationality. If I don't believe — *know* — that I will survive, I will not dare to jump but instead make do with minimal improvements to other people's mistakes.

Because there is much at stake here! The solution might not exist. We might be lost. We might be wasting our time wandering around in a dark forest of self-delusion and unfulfilled aspirations. We probably *are* lost! What arrogance to think otherwise. But to think that is to give up. Without belief, hope, self-delusion — call it what you want — without that, I don't know how to fly.

And isn't this what research is all about? To risk it all, *all of it*, to be arrogant and defiant and *never* surrender. To keep going over those big waterfalls — and to always choose the biggest! — trying to imagine the solution before you hit the bottom; before the rocks destroy you, which they probably will. But then you'll at least have tried.

When my intellectual kayak goes from horizontal to vertical, I am always faced with this question: *dare I spend my entire life pursuing one single idea?* When is it time to quit? When is it time to admit failure and mediocrity? How will I know?

The answer is that I'll never know. That is what makes it so scary.

But this time I did find what I was looking for, the conjugate variables. One day during my hike through the Brenta massif I hit the bottom of this waterfall and I found sweet water and no rocks: I finally got an intuition about how these variables should be included in our work.

The conjugate variables should interact with the holonomy-diffeomorphisms (the operators that move stuff around) in a fashion that reproduces the canonical commutation relations of a quantum gauge theory. We knew this and the idea I had was simply that they should come from an elementary operation that changes the way holonomy-diffeomorphisms work. When you move *stuff* from one region of space to another there are multiple, even infinite, ways of doing it, and the conjugate variables should communicate between these different possibilities.

I was excited to get back to Johannes and tell him about my idea. I arrived at his place early one morning after taking the night train from Munich and sat in his kitchen drinking tea with him, feeling impatient to get through our standard ritual of pleasantries to tell him about my thoughts. But ritual is ritual, so we talked about his teaching, my hike, politics and the weather in Italy and Germany and global warming too, and finally, when we had run out of alternative topics, I opened the door to the second part of our exchange, our work.

I told him that I thought I had an idea about the conjugate variables and then I paused to let him speak. Johannes then told me that he too had an idea and explained it to me; it turned out to be precisely my idea, just much clearer and better formulated.

A sting of disappointment went through me. I would have liked to have been there first, to have the joy of telling Johannes about the treasure I had found. But these feelings quickly faded, and instead, I felt excited. It was obvious to me that we now had the algebra that we had been searching for over the last decade. A foundation. This was it. With this algebra we would either make it or break it — there would not be another attempt from our side.

And the structure seemed so obvious, I couldn't believe that nobody had looked at this before us. And why hadn't *we* found it sooner? We decided to search the literature to see if it really could be true that this algebra was novel to the mathematical community.

✡

The first question that arises once the QHD algebra has been found is whether a quantum theory can be built over it. Does it have a Hilbert space to live in? Not all algebras do and so we needed to determine whether it does.

We spent 3 years analyzing this question and we are now convinced that the answer is yes. A Hilbert space does exist and so we have a quantum theory at hand — what we call *Quantum Holonomy Theory*.

So, what does this theory look like?

Well, first of all, it looks *young*. We have just found it and are still getting acquainted with it. But there are a few important features, which begin to stand out.

Perhaps the most striking feature — and certainly the one that has surprised us the most — is that what we have found is *not* a theory of quantum gravity. The theory offers a novel solution to the one key problem that *any* theory of quantum gravity must solve, namely the problem of screening off distances shorter than the Planck scale, as discussed in chapter 2. But it does so in a manner that does *not* involve a quantization of the gravitational field.

In other words, it makes a theory of quantum gravity obsolete.

I find a strange beauty in this: we have been searching for a theory of quantum gravity for so many years and then we end up finding something completely *different*. This is the nature of adventure; you never know what you will find and sometimes you don't even know exactly what you are searching for. I can't help thinking that this is the perfect solution to a problem, that has remained unsolved for decades: that the object everyone was looking for simply does not exist.

Quantum holonomy theory is a theory of quantized gauge fields, but the way these gauge fields are quantized does *not* match the prescriptions of a theory of quantum gravity. It does involve gravity, but gravity remains classical.

There is another remarkable feature, which is related to what I have just said and that is that the Hilbert space in our theory is intrinsically *non-local*.

Non-local? What does that mean?

Locality is a key ingredient in all the physical theories that we know of today. Einstein's theory of relativity and the standard model of particle physics are both local, which essentially means that they involve quantities where numbers or operators are assigned to *points* in space (and time) — they are based on either *fields* or *quantum fields* — and these fields interact point-wise.

Take Einstein's theory of relativity, where the basic ingredient is the metric field. There is nothing in Einstein's theory that prevents this field from having highly localized configurations such as black holes, where the metric field diverges in a point; or the big bang, where everything is collapsed into a single point. Einstein's theory is *local*.

It turns out that quantum holonomy theory does not have this feature. It is *not* local. In this theory, it is *impossible* to localize anything to a single point. The Hilbert space — which, if you recall, is the place where operators act to produce the numbers we hope to measure in experiments — assigns zero weight or probability to configurations that are localized in single points. In a way, you can say that single points do not exist in this theory.

And interestingly: as far as we know it is *impossible* to construct a non-trivial Hilbert space for the QHD algebra, which involves locality. It seems that the QHD algebra carries non-locality as an innate property.

If you agreed with the simple argument I made at the beginning of this book as to why a final theory must exist — an argument that combined quantum mechanics with Einstein's theory of relativity and which concluded that single points are operationally meaningless because they cannot be measured — then you might appreciate that this non-local feature of quantum holonomy theory is *precisely* what one would expect from a fundamental theory.

— *"Okay, so this sounds good, but what does it really mean?"* you might ask.

Good question. We are still in the process of analyzing all this but let me give you an idea about where we think this non-locality is taking us.

First of all, we believe that this implies that the big bang was not a bang but rather a bounce. If the fundamental theory does not allow localization in a single point then there is no room for a big bang where everything, the entire universe, is collapsed into a singularity, an extremely local event. This leaves only one possibility, namely that of a bounce, where an earlier universe collapses into a near-singularity for then to bounce into the universe, which we now so happily inhabit.

Likewise, we also believe that this theory implies that black holes must have a bottom. In Einstein's theory of relativity black holes involve a singularity, where the curvature of space-time becomes infinite. But according to quantum holonomy theory, such a singularity cannot occur. The singularity within a black hole will never be formed. We think.

There are other aspects of this non-locality that are deeply interesting — it involves what we call a dynamical regularization, which means that the fashion in which scales beyond the Planck scale are screened off will be subject to time-evolution — but this is something that we are still analyzing so I cannot say much about it just yet.

✡

The principle that we propose as final — and which gives rise to quantum holonomy theory — has a touch of the absurd to it. The idea that the scientific search for truth, which the ancient Greeks set in motion and which thousands and thousands of scientists have continued throughout the centuries, could find its conclusion with something as simple as an algebra of 'moving stuff around' seems almost blasphemous. *"What? Is that it?"* one might think. *"Surely, Nature could have thought of something more interesting than that?"*

The QHD algebra is strangely void of conceptual *depth*: the mathematics of moving stuff around in space does not tell us anything of existential value nor does it tell us anything deep about the Universe (other than it may have chosen a near trivial mathematical principle as its foundation). It seems so eerily ordinary.

But what did/do we expect to find?

The British physicist Steven Hawking wrote in 'A Brief History of Time' that

> *"... if we do discover a complete theory, it should in time be understandable in broad principle by everyone, not just a few scientists. Then we shall all, philosophers, scientists, and just ordinary people, be able to take part in the discussion of the question of why it is that we and the universe exist. If we find the answer to that, it would be the ultimate triumph of human reason — for then we would know the mind of God."*

Hawking expected a final theory to tell us something of existential value, he clearly did not anticipate a theory that is conceptually empty.

It is apparent that most physicists hope to find something of profound value at the foot of the scientific ladder, something that will not only bring deep intellectual insights but may even tell us something about reality that goes beyond the strictly scientific. Many physicists would deny that they expect to find something of existential value in a final theory, but that does not imply that this is not part of their motivation for taking part in the search — in fact, I'm convinced that it is.

But the principle that we propose does not seem to offer any of that. Instead, it has the form of an almost self-evident triviality, which one might as well have thought of right at the beginning, except that the principle that we suggest does have general relativity as a prerequisite, so nobody before Einstein (or, at least, before Riemann) could have come up with it.

In any case, if this principle should turn out to be correct and Nature has indeed chosen this option, then I think that we will find ourselves in a Shell Beach scenario. That is, a situation where the place that we have been searching for turns out to be empty. As if the search itself dissolves into thin air.

But I ask again: what did (or do) we hope to find? What type of theory could have met (or meet) our expectations?

Imagine if in a profound way the final theory entails a proof of the Riemann hypothesis (a famous and unsolved problem concerning the growth of prime numbers first posed by Riemann) and links all the major research directions in modern mathematics in a hitherto unseen way. Or imagine if the final theory entails deep insights into apparently unrelated research fields such as the study of consciousness and the origin of life. Would that satisfy us? Yes, absolutely, that would be the kind of theory that would meet our expectations. Thank you very much! And why? Because it would tell us something about reality that we didn't know before, something of physical, philosophical, and existential significance. And precisely because such a theory would not leave us empty-handed it could even postpose the end of the search by opening up new directions of research. Some of the mystery would remain.

And I think that this is the key: If we find something truly amazing, our sense of wonder and mystery will be preserved. The truth is that we want to know *without* losing our sense of awe, without losing the mysterious feeling of *not* knowing. And that is impossible.

In the end, however, I do believe that the Shell Beach scenario poses the most interesting outcome and, as already said, the most likely. The more structure and content a candidate for a final principle and final theory has, the more exposed it will be to further scientific reduction. It certainly seems plausible that a final principle could include deep mathematical insights, but to imagine it capable of reaching out to other scientific disciplines — outside physics and mathematics — seems far-fetched. And if the choice stands between a principle that is conceptually *almost* empty and one that is *completely* empty, then I think that the latter is by far the most compelling and beautiful. Just as zero is the perfect number so must the perfect endpoint of the scientific search be the empty one.

CHAPTER 21.

ASHES FALLING FROM THE SKY

"Man stands face to face with the irrational. He feels within him his longing for happiness and for reason. The absurd is born of this confrontation between the human need and the unreasonable silence of the world."

— **Albert Camus**

It's my last day at the Niels Bohr Institute. The 31st of August 2011.

I'm sitting in my office with the door closed, looking out the window. It's four in the afternoon. I'm waiting.

Often when people leave this place, they buy snacks and wine and have a small reception for the group to mark the occasion. I don't feel like doing that. Instead, I've spent the day clearing the office, sorting out my papers, and throwing away those that I will not need.

I've wiped the blackboard clean and emptied the wastebasket and now I'm sitting with my feet on the table looking out the window, waiting for the last hour to pass.

When it's half-past five I get up. I close my computer and shove it into my shoulder bag. I get my jacket and pause for a moment by the door, I stand in silence and listen — and then I open it. The institute is quiet. I walk out, close the door behind me, and lock it.

I walk down the stairs, put my office keys into an envelope and shove it under the door to the secretary's office. Then I walk down to the entrance and leave the building.

I don't meet anyone on my way out. I walk across the yard under the white puffy cloud of a blossoming cherry tree and disappear into the city.

✡

March 2012

Ashes are falling from the sky.

Little black pieces, like burnt paper, unreadable.

Every minute, like the hand of an unsteady clock, sometimes early, sometimes late, but unrelentingly ticking time forwards, black pieces of ash fall from the sky.

It is as if the clouds themselves have been burnt, as though they all ignited and combusted and vanished in a brief and bright flare leaving only this silent drizzle of ashes falling back to earth.

It's close to noon and the sun is barely visible. I can see it high in the sky, like a perfect disc in a sea of grey. I watch it for a long while without moving my eyes, I look into the sun with a kind of defiance, as though it has finally been defeated and I celebrate its demise with a victorious stare, eyes wide open.

I'm sitting on the hood of the large Toyota engine in the rear of the longboat and I can feel the vibrations of the powerful beast just inches beneath the wooden planks. Every time we go up a series of rapids the engine roars wildly and the vibrations intensify.

We are passing through a burnt landscape. It's like a war zone. I see nothing but burnt lands, trees, fields. Black, all is black. With every turn of the river, I hope to see lush rainforest, green branches reaching out over the river as a show of Nature's inability to contain its immense joy in itself, but with every turn, I see ever more burnt land. It is as if the whole country has been burnt and that nothing but black remains.

I'm in Laos, going up the Nam Ou river where I am hoping to find a quiet refuge near the Chinese border. The farmers are using the slash-and-burn technique and now they are burning. Everywhere. The sky, which I imagine to be clear blue today, is grey, like twilight at noon.

We're a handful of foreigners who have hired the boat to travel up this river to reach the village of Nong Khiaw. I don't know any of the others, we're all strangers in a strange country. Once in a while, we pass some Chinese gold panners, who stand in the river with their broad hats and flat pans. Some of them laugh when they see us, others just stand and stare.

I've made a stopover on my way back from Canberra, Australia, where I visited Adam, the mathematician. I left most of my stuff in a hotel in Bangkok and crossed the border into Laos. There is no hurry to get back home, being unemployed can wait.

I catch a piece of ash that falls in front of me with my cupped hand and I try to read it to see if some of the words are still decipherable, to see if there is a message for me to take home from all of this. Words from the past, spoken and unspoken, they now return as silent witnesses of everything long forgotten.

I think that we need these rivers, we need them to journey upstream in search of the source, the origin, the place for answers. We search for the fire that caused this rain of ash that we walk in. And even if these rivers have many springs, we need to find just one of them, because we need to find a feeling, an idea, we need to find a shadow of an answer.

The river is narrowing — and the water is getting shallow. Sometimes we have to get out to push the boat over a stretch with loose gravel; a group of foreigners in a foreign country pushing a boat upstream. I wonder if these shallow waters mean that we're getting close to our destination.

January 2013

I've found my refuge.

Two of my friends, Asger and Pernille, wrote about a luxury lodge in the mountains of northern Vietnam. Come and visit, they wrote and told me that they now manage the lodge for a tourist company. You'll love it here, they continued, stay as long as you like.

January and February are the quiet months here. It's winter, it's cold, it's often misty, and the lodge, which is situated on a beautiful mountain hilltop above a steep valley in a remote area of the Black H'mong and Red Dao minorities, has very few visitors. Like, none.

They needed to go away for a while, Asger and Pernille, and they asked me to come because they knew that I was looking for a quiet place to think, a place to just be for a while, and a place to work. And they asked me because their little kitten, Henry — which they have just taken in, hoping that it will later grow into a tiger that can clear their house of the vicious Vietnamese rats that roam the house at night and that once bit Pernille in her sleep — needed someone to look after him.

So here I am, in Northern Vietnam, alone in a fabulous lodge with this little yellow furry friend, who immediately took me for his mother and who is just as scared of the rats as I am.

I've just spoken to Johannes on Skype. We published two papers on the holonomy-diffeomorphism algebra half a year ago and now we're trying to figure out how to include the other variables — the *conjugate* variables — in the construction. There is a problem, which we have discussed for a long time — it has to do with the Hilbert space.

We talk every week, sometimes every second week, via Skype. Sometimes I'm in Denmark, sometimes I'm away — on an island in Thailand or somewhere in India or visiting friends in Canada — it makes no difference, the internet is everywhere, even here on a remote mountaintop in a part of Vietnam that is so far north that it might as well be China. Our conversation is one of the few constants in my fluid life, it slowly moves forward, one tick every week, where we tell each other what we have been thinking and where we're moving.

And we are moving. I often use a mental exercise, where I think about where we were exactly one year ago. What did we discuss this month last year? What problem were we working on? It helps me see the progress, that it all makes sense. It helps me see that it's not just me that can't face the fact that my career is dead, the fact that I've failed and didn't manage to break through the academic walls, it helps me see that I'm not fooling myself, that our progress is real, and the direction of our work makes sense.

It's like a fixed point, this mental exercise, or a fixed star or just a fix, it gives me courage. I can still see it and it doesn't matter if nobody else does.

Today I took Henry out in the field outside the house. He's very new to this world. Until today he had never been outside; his universe was the house with the bamboo walls; the sleeping room with the bed and the mosquito net; the large room with the TV and the kitchen, and the bathroom, where he stands outside the door and screams if I go in without letting him come with me.

Until this day, when I took him in my arms and walked out in the middle of the field with the high grass. I sat down on a bench and put him down on the ground next to it. It was a wonderful experience to see him take it all in. He was so alert, so very curious and so very careful. He slowly walked a few meters away from the bench and came racing back when a cricket suddenly jumped up in front of him.

So full of joy.

The world is new, it always has been and all it takes to prove it is a little yellow kitten.

CHAPTER 22.

TIME MATTERS

The seconds pour
over the edge
and fall
like flakes of snow
spread in the wind
unable to form
minutes
hours
days
years
a life.

And yet
the snow falls
and forms
a surface
pure, untouched,
of time waiting
for tracks.

— J.M.G.

In chapter 4 I wrote that a quantum theory consists of two halves: an *algebra* and a *Hilbert space*. This is not quite the full story. There is one more piece that is needed to complete the picture, and that piece is *time*.

Once you have found a Hilbert space for your algebra of operators to live in you need to define a *dynamical principle* that determines how your theory will evolve with time.

Or rather, you have to *make* time happen.

In physics, we say that the algebra and the Hilbert space form the *kinematics* of the quantum theory. What we are talking about now is the *dynamics*.

Normally a dynamical principle is given by an operator called the *Hamilton operator*, which is required to satisfy a few basic mathematical conditions, and which then determines the time evolution of states in the Hilbert space.

That is, given a state at a certain time, the Hamilton operator will tell you what that state looks like at a later time.

One crucial condition that the Hamilton operator is required to satisfy is to reproduce the Hamilton of the corresponding *classical* system in a classical limit. This requirement implies that the form of the Hamilton operator will, to a certain degree, be determined by the classical theory. This means that we are once more faced with the question that we discussed in chapter 19 as to what is primary, the quantum or the classical? If the dynamical principle is dictated partly or completely by the classical theory, then the quantum realm *cannot* be viewed as primary: the operational arrows go from the classical to the quantum, not the other way around.

However, another possibility exists. An alternate way to obtain a dynamical principle, where the operational arrows are solidly rooted in the quantum realm and where the classical theory does *not* dictate the form of the Hamilton operator. What I am talking about is an application of the toolbox of noncommutative geometry.

But before I venture forward with my explanation, let me offer you yet another friendly warning: what comes next will sound like science fiction of the crazy kind. But no matter how crazy this may sound I can assure you that the mathematics behind it is perfectly solid. So, let's buckle up and get on with it!

Let us begin by once more reminding ourselves that Einstein's theory of relativity tells us that gravity should be understood in terms of *curvature* of space and time. The reality in which we live has a complicated geometry, it has curvature.

Now, think about your living room. Your sofa may stand in front of the kitchen wall with the table placed in front of the sofa. Or perhaps you have the sofa located on the other side of the table? Or perhaps your sofa stands on top of the table (why not?); heck, you may even have all your furniture standing upside down. There are obviously lots of different ways in which you can arrange your living room. Now, imagine the abstract space, where each *point* in it is one particular way of arranging your living room. This is an enormously large space, even infinite-dimensional. And imagine that this space, which consists of all the possible ways of arranging your living room, has *curvature*.

This sounds crazy, I know, but in a certain sense, this is precisely what Johannes and I suggest. The 'living room' that we are considering is the space of all possible ways a gauge field can be configured. Take for instance electro-magnetism, where we have the electromagnetic fields. Here we can think of the huge space where each point is one particular configuration of the electromagnetic fields.

This means that there will be one point in this space that corresponds to the sun shining (electro-magnetic radiation), and another point will be the configuration where the sun does not shine. One point in this space will be a configuration with radio waves and Wi-Fi, another point will be a configuration without all that.

And what we have found is that it is possible to formulate a geometry *on* this infinite-dimensional space of all possible field configurations. We have discovered a 'relativity theory' *over* the space of all possible scenarios. And this construction employs precisely the mathematics of noncommutative geometry discussed in chapter 15.

The idea that we propose is that the operator that gives us the curvature of this infinite-dimensional space is identical to the operator that gives us time evolution in our theory. The Hamilton operator.

The beautiful thing about this idea is that this operator is not arbitrary. Its form is, to a large extent, predetermined. There is essentially only one way to formulate it and that means that we have a theory where the operational arrows clearly run from the quantum towards the classical. The quantum is primary.

✡

When Johannes and I visited Penn State University in 2009 to meet professor Abhay Ashtekar we also visited the mathematician Nigel Higson, whom I had met earlier at various conferences. Nigel Higson is a professor at the Institute of Mathematics at Penn State.

Throughout the years I have met many brilliant mathematicians, some of the best in the world, I believe. It is my experience that most of them conduct themselves in a very controlled and proper fashion. They have a very *conscious* way of being in the world. It is as if they see things in a different and stronger light, as if they swim in a clear pool of pure logic. It is not all mathematicians who are like this, I have also met some very impulsive and messy thinkers and they were not the least talented. But it is my experience that many of the best mathematicians have this very deliberate and lucid quality and Nigel Higson is one of them. He is highly intelligent and always expresses himself in a very precise manner. His publications are equally precise — the parts that I am able to read seem almost easy, like a broad path through the jungle. Higson doesn't just cut the largest branches to then have his readers stumble on the smaller ones; he clears a broad path, which makes the walk truly pleasurable. His lectures have the same style. It is always inspiring to watch this kind person demonstrate what logical and rational thinking looks like.

During our visit at Penn State, Higson asked us to come by his office one day where he told us that he had read our papers and that he liked our ideas but that one of the key components in our construction could not be the right one, in his opinion.

Higson was right, both Johannes and I had already realized that something was missing, but at that time we did not know what else to do. Not until eight years later, in 2017, when we published two preprints with a proof that our Hilbert space representation exists in a strict mathematical sense.

During those years I had kept Higson's words in the back of my mind and now that I believed that I knew the answer to his question I was eager to write to him, and so I did. I sent him an email one Sunday morning and Higson replied just six hours later.

In his reply, Higson wrote that the solution we had found reminded him of a mathematical construction, which he had formulated with the mathematician Gennadi Kasparov in 2001, and sure enough, when I read their paper, I realized that he was right. One of the key ingredients in our construction — the Hilbert space — had the same structure that Higson and Kasparov had used 16 years earlier.

But there was something else. One of the key components in their construction looked different from the one we were using. Theirs had an extra little 'thingy' attached to it, a technical subtlety, which from a mathematical perspective made a lot of sense but whose physical significance was not clear to us.

But it soon became clear. After having spent some time thinking about it, we realized that the operator that Higson and Kasparov used was precisely the operator we needed in order to generate time. To get the Hamilton operator.

Higson's and Kasparov's operator is a curvature operator, so here is what we propose: that time is generated by an operator that defines the *geometry* of the abstract space of *all the possible ways in which the world can be arranged.*

I am convinced that Spock would have loved this idea.

✡

It took us three years to work out the details of this new operator. There are some mathematical constraints that required us to modify the operator that Higson and Kasparov had used before we could fully implement it in our framework. But in August 2020 we finally nailed it.

For historical reasons, we call this operator a *Bott-Dirac operator.*

The Bott-Dirac operator is extremely interesting. As I said, it defines a geometry on the huge space of all possible ways a gauge field can be configured. But it also has some other spectacular features.

First of all, it naturally involves the correct time evolution for not only a quantized gauge field (in technical terms it entails the Hamilton operator of a Yang-Mills theory, which is a theory of gauge fields), but also of matter fields. Matter fields are what we call fermions and our Bott-Dirac operator automatically introduces quantized fermions into our mathematical machinery.

This is what I find so remarkable: when we construct this operator, which gives us our geometry, then we are *automatically* led to a theory that involves quantized matter.

This means that matter plays an intrinsically geometrical role in this framework. It is the geometry of the huge space of all possible field configurations that produces *stuff*.

I find this incredibly interesting.

✡

The application of the Bott-Dirac operator invites a particularly interesting consideration concerning the origin of Einstein's theory of relativity. A key ingredient in both the special and the general theories of relativity is the *signature* of the metric field. For general relativity to be what it is, the metric field must have what is known as a *Minkowski signature*. This mathematical ingredient encodes the very essence of relativity theory and in particular tells us that the speed of light is constant and the same for all observers.

But here is the thing, the metric field could have other signatures than Minkowski. It can be argued that a more natural choice would be a Euclidean signature — a situation where Einstein's theory would not have all its odd features. Or it could have a signature that includes very exotic situations such as two time dimensions and two spatial dimensions. But Nature didn't choose that, it chose a Minkowski signature.

Now, the signature of the metric field is encoded in the Hamilton operator. It can be read off from it, so to speak, and that means that when the Hamilton operator can be obtained intrinsically by employing the machinery of non-commutative geometry, then the signature itself will be a derived quantity. It will be an *output*. Then we will have an answer to the question: why relativity?

It will be innate, there will be no choice. This will be precisely the kind of emergence one would expect from a final theory, where the operational arrows take their point of departure in the quantum realm and where the classical world is truly secondary, emergent, and approximative.

Can this work? Could the Bott-Dirac operator be exactly the right operator to provide the world with what it needs the most, namely time? Could this theory of ours really be a theory that describes the real world out there? Could we be right?

We don't know yet — but we are determined to find out. What we do know is that the simple *ansatz* of moving stuff around in space has led us to a theory that involves essentially *all* the key ingredients found in the two major theories of fundamental physics, general relativity and the standard model of particle physics: quantized gauge fields; quantized matter fields; curved background; and an algebraic structure that strongly resembles that of the standard model. Whether these ingredients eventually combine to produce exactly the dish that we intended to cook, we don't know. We are hopeful but being hopeful is worth very little in this game. What we need is proof.

CHAPTER 23.

AFTER A FINAL THEORY

„a purely intellectual worldview completely without mysticism is an absurdity.“

— Erwin R. J. A. Schrödinger

In the movie "Dark City", when John Murdoch eventually finds Shell Beach, he finds an empty void. Literally, he knocks down a wall and is faced with empty space. Where he hoped to find answers, a foundation, he finds nothing at all.

The movie ends with John Murdoch creating Shell Beach himself. He creates the beach, the ocean, the sky, and everything else. He knows that no reliable memories exist, that the inhabitants of the city will never know where they came from or who they are, and his reaction is to take matters into his own hands and create reality himself, to his liking. The empty void that he finds sets him free.

It is an interesting question in which way the possible discovery of a final theory will affect we humans and our civilization. Imagine if we succeed and we find that theory. What will it do to us?

All great scientific discoveries impact societies in ways that greatly surpass the boundaries and borders of their specific scientific disciplines. Insights morph into metaphor and from metaphor it is only a tiny leap to myth. And metaphor and myth are what runs in the veins of any civilization. I am convinced that the discovery of a final theory will have a momentous impact on our civilization. So, let us ask again: what will the discovery of such a theory do to us?

To search for an answer to this question I would like us to first look back and consider briefly the history of western civilization. What do we see? To a large degree, we see a history of conquest, a history of growth, and a history of expansion. Through the past centuries, western man[7] discovered the Americas, the Orient, Africa, the entire globe — and wherever he went he found territories, whose conquest brought him wealth beyond imagination.

All this changed — in my opinion — on July 20th, 1969 when the American Apollo mission put a man on the Moon.

The conquest of the moon was in so many ways a logical next step. There was no more territory on earth for man to conqueror, it had all been explored and so space had to be the place to go looking for the next conquest.

And of course, the Apollo missions were a huge success and widely celebrated as yet another triumph of humankind. But the truth remains that what we found up there was not another fertile continent with wealth

---

[7] The term 'western man' should here be understood both literally as a reference to the fact that the history of the western civilisation has to a very large extend been dominated by men and nonliterally as a reference to the masculine element in the human psyche.

beyond imagination. There is a good reason why the Apollo missions were not followed by a series of new missions to the moon with colonization, exploration and exploitation: there is nothing up there for us, it's just a piece of rock (there may be important mining for future generations, but that is another story), and worse: when we look further out into space we see nothing but similarly barren rocks, which offer depressingly little in terms of new territory — anything of real interest is ridiculously far away.

The truth is that the Apollo missions marked the end of the era of man as conqueror. There simply is no more territory for us to conquer.

We are stuck on earth and we will never leave it. This is it, there is no other place for us to go.

Parallel to the conquest of physical territory the history of western civilization chronicles another kind of conquest, namely the *intellectual* conquest of the external world — *science* — where man has conquered new intellectual territories by probing ever deeper into the fabric of Nature. An expansion of the mind.

This expansion has three directions, three natural *doors* to reality, which are the *very small*, the *very large,* and *consciousness*. These are the natural places to look for fundamental truth. There is the reductionist approach that takes us from biology to chemistry and through physics and beyond; there is the approach of astronomy and cosmology, where we look out into space; and there is finally the looking itself: the observer of it all, ourselves, the subjective element in the Universe. Science has explored all three doors, most extensively the first two, and we now know that the exploration of the very small and the very large ultimately meet up in the same place and the same question, namely the question of a theory of quantum gravity and the question of a final theory.

There is, of course, a fourth door, the ultimate one, death. We'll all walk through that one and then we'll know. Or we won't know. But we won't know which one it is until we've closed it behind us and so, for now, this door is not a door for us to reckon with.

This process of intellectual conquest will end with the discovery of a final theory[8]. If it exists, it will be the end of the road and mark the definitive end of the era of man as conqueror of the external world — physically *and* intellectually.

And if what we find is a Shell Beach scenario, where the final theory does not tell us anything of existential significance, then it will be like the Apollo missions to the moon — an incredible, unprecedented success with a deeply tragic core. We will have searched everywhere and found the end of the road, but its discovery will leave us hungry and with no place else to go.

What we had thought to be a door — a gateway to ultimate truth — will then have turned out to be a barrier. A definite, insurmountable, and absolute barrier.

And here is the question that begs to be answered: what will happen to man as a conqueror when there is nothing left to conquer? Not in the physical world and not in the intellectual realm?

---

[8] Of course, exploration of already conquered territory — such as atomic, nuclear and particle physics, chemistry and biology — will continue and new discoveries will be made. The point here is that we will not open *new* intellectual territories beyond the final theory.

To answer this question, we need to understand why conquest has been such an integral part of western civilization. Not all civilizations have been like this. In fact, for most civilizations, expansion was not a part of their cultural DNA.

I find a trace of an answer to this question in the autobiography of the psychiatrist C. G. Jung, in which he describes a meeting he had with a Pueblo Indian chief named Ochwiay Biano, who described the white man to Jung:

> *"... their eyes have a staring expression; they are always seeking something. What are they seeking? The whites always want something; they are always uneasy and restless. We do not know what they want. We do not understand them. We think that they are mad."*
>
> *I asked him why he thought the whites were all mad.*
>
> *"They say that they think with their heads," he replied.*
>
> *"Why of course. What do you think with?" I asked him in surprise.*
>
> *"We think here," he said, indicating his heart.*

Perhaps the Pueblo chief was right. Perhaps we, as a civilization that thinks with our heads and not with our hearts, are mad, and perhaps this madness is caused by an absence of something essential that we lost many generations ago, a spiritual dimension in our being in the world. Could it be that what we search for in the external world and in our obsession with nailing it all down in closed formulas, is something within ourselves, something that the Pueblo Indians and many other so-called primitive people, including, presumably, our own distant ancestors, possessed?

Perhaps what we are searching for in the outside world is something we have lost within ourselves?

The discovery of a final theory will be celebrated as the greatest scientific discovery of humankind ever — and rightly so. It will be an incredible intellectual triumph, an unprecedented milestone in the history of humankind and the immediate reaction to such a discovery will be one of jubilant celebration.

But the immediate stage of celebration must at some point be followed by a period of crisis: So we found the final theory, the ultimate scientific truth, and what does it tell us? Some empty mathematical principle (whatever it may be, holonomies or something else), what does that actually tell us? *Very little* or *nothing at all*. And even worse, the discovery of a theory that beyond any reasonable doubt really is final will deprive us of the possibility of a continued search.

What will this do to us? How will we react?

It is difficult to imagine precisely how the discovery of a final theory will affect society other than it must lead to a kind of intellectual and existential crisis triggered by the realization that there will be *no other place in the external world to search for truth*. Where are we going to search for answers once the scientific project has ended? One thing is to be skeptical about what answers such a search may produce — and most scientists I know would express such skepticism — it is an entirely different situation to actually *see* that it produces no answers at all. In the first case we have at least the task of searching, but what do we have once the search has ended?

At the core of the scientific project lie dormant issues of mythological proportions. The scientific project grew out of the ancient mythologies, it replaced them and made them irrelevant and thus within the scientific project is the — mostly unspoken and partly unconscious — hope that in the end it will provide us with some of the answers that the mythologies used to give us, answers concerning our relationship to the Universe. Stephen Hawking put this into words when he said that we shall *"know the mind of God"*. If the scientific project comes to an end *without* meeting this need (and how could it end in any other way?) then it will leave us in an intellectual and existential crisis.

In psychoanalysis, a crisis is seen as a possibility of change, where the old must die for the new to be born. And in many eastern religions, the entrance to the most sacred part of a temple is guarded by terrifying threshold guardian figures symbolizing the painful process of attaining the highest spiritual wisdom. Likewise, one could speculate, and even hope, that the crisis caused by discovering an empty final theory will set in motion a gradual reorientation of western psychology away from conquest and towards something new — perhaps a way of being in the world akin to that of the Pueblo Indians.

Personally, the thought of discovering a final theory that tells us little of consequence, fills me with a feeling of emptiness, a quiet sadness. There is an air of failure over such a discovery. As if the whole scientific project is a dead end. It is a lonely sadness — not the sadness of losing something but the sadness of never having possessed it — a feeling of being profoundly lost. There is something claustrophobic about this thought too, a desperate and deeply disturbing feeling, like being enclosed alone in a box with impenetrable walls. Our box is the Universe and like detective Walenski in "Dark City" we will say *"there is no way out, we have tried"*. Regardless of whether Johannes' and my work turns out to be correct or not, I can't help feeling that we are in the business of closing, not opening, doors and that our work does not bring any message except the obvious one: *go search for truth elsewhere*.

This truly is the domain of the absurd as Camus described it: it is not Nature on its own, it is not the intellect of Man by itself, it is the *combination* of the two that is absurd, where the former cannot provide what the latter desperately seeks, namely *meaning*. Camus writes in 'the myth of Sisyphus' that

> *"if man realized that the universe like him*
> *can love and suffer, he would be reconciled"*

— something that is certainly beyond the reach of physics and science in general.

Despite all this — or perhaps precisely because of it — I believe that the most interesting question is what might follow this initial period of crisis. The reaction that John Murdoch had when he discovered that Shell Beach is nothing but an empty void was to take matters in his own hands. *It set him free*. I think that this is an interesting question: whether the discovery of a final theory that tells us nothing of real interest could set us free in an existential and perhaps spiritual sense.

According to the American scholar Joseph Campbell, any mythology must include four basic functions:

1.   **A mystical function:** a primary function of a mythology is to open up for experiences of a transcendental character. Mystical experiences, which establish a meaningful relationship between the individual and the Universe. A sense of purpose and belonging.

2.   **A scientific function:** a mythology should explain reality — why it is this way and not another way. And it must be in accord with the reality that people experience now and not how people saw it in some other historical period.

3.   **A sociological function:** mythologies work to uphold the social structure of society. By giving people a sense of purpose in their particular role and their lives they serve as a conservative force in society.

4.   **A pedagogical function:** a key function of a living mythology is to provide guidance to people in their individual lives as they pass through the various stages of life, from birth to death. A mythology must guide the individuals in their transition between the different stages of life.

All civilizations throughout history have developed mythologies, we are the first to live without one. It seems, however, that we are born with an innate religious or spiritual capacity and that the mythologies are there to activate these capacities. Science is the closest we come to a mythology in our culture at the present moment in history, but except for the scientific function, science does not meet Campbell's requirements. It does not give us access to the mystical experiences, which people in previous cultures had — and which people in some cultures still have — in fact, it probably prevents us from having them by communicating a worldview, in which we are seen as a random occurrence. And science does not help uphold the structure of society nor does it provide any guidance whatsoever in our personal lives.

The question is whether a society in a post-final-theory world will develop a new kind of mythology? After all, the formation of mythologies seems to be deeply rooted in humankind. The question is whether a final theory that does not tell us anything of existential significance could even be beneficial for the emergence of a modern mythology?

Clearly, a new mythology must be in perfect agreement with science as we know it — any discrepancy will delegitimize it — but since most of the basic functions of a mythology are concerned with our inner lives, our spiritual lives, and *not* with the physical reality that we experience, adherence to scientific truth does not seem to pose a serious obstacle for a modern mythology to emerge. The problems with the old mythologies are simply that they were formulated several millennia ago in a time when science had not yet been developed. The exact details or time of the creation of earth is not essential for, say, the Bible and the Christian gospel, what truly matters is not physics or even metaphysics, but how the words and images of these texts *affect* us. The crux of a mythology is not whether it is true in a scientific sense but whether its symbolic language speaks to those parts of our psyche, where our religious or spiritual potentialities lie dormant. Mythologies do not speak to our rational faculties, their language is the language of poetry. It is the language of the heart, and what matters is not whether their stories make sense to the intellect but whether they speak to those other parts of our being.

When we lose the implicit understanding that mythologies speak in a language of metaphor, where the actual content of the stories plays a secondary role to the primary function of establishing a certain state of mind in the individual, then the stories of the old mythologies, which are often in direct conflict with what we today *know* to be true, no longer work on us. It might have made sense 2000 years ago that God is in heaven (as in, up in the sky) but today we travel through the skies, we know of Einstein's theory of relativity and therefore these metaphors (because that is of course what they are — both 'heaven' and 'God' are metaphors) lose their power. Instead of speaking to us at a spiritual level, we reject the entire story on scientific grounds. It leaves us cold.

The fact that the language of mythology and religion is primarily symbolic and should not be read literally is very often misunderstood by scientists, and quite frankly, also by many (or even most?) religious people. This is one of the main sources of all the tedious misunderstandings that often characterize discussions on science and religion. If this fact were widely recognized, it would provide a basis for a much more interesting exchange.

But the point here is that except for the second, scientific function mentioned above, the domains of mythology and science are strictly separate: the first lies somewhere between the metaphysical and our inner spiritual lives — with a strong tilt towards the latter — whereas the domain of science is physical reality. As long as the boundaries between these domains are not violated there should not be a problem formulating a new mythology. Science has — per definition — *nothing* to say about metaphysics, and thus we are free to construct whatever stories we like, whatever stories can *help us* in our daily lives, as long as they are not in conflict with the science that we know. And the question must be whether such stories, as a central part of our collective psychology, are not *bound* to emerge at some point as a new mythology?

I have no idea what a modern mythology might look like, or whether it will ever emerge, but as already said I suspect that the discovery of a final theory, which tells us nothing of existential value, will actually make it easier for it to happen. It seems to me that such a theory will leave more room for *artistic* interpretation — like an empty canvas, upon which you can paint whatever you like as long as you stay within the boundaries of the frame — and that it may even provide the sense of mystery that a mythology feeds on.

In fact, I can't help thinking that in an odd way the possible emptiness of a final theory will, eventually, turn out to have a mythological character in itself. There is something mystical about such an endpoint. It leaves us with an even bigger mystery, which now lies outside the domain of science. Perhaps this is what Hawking referred to when he wrote that

> "*we shall all, philosophers, scientists, and
> just ordinary people, be able to take part
> in the discussion of the question of why it
> is that we and the universe exist.*"

The discussion that Hawking referred to is a discussion that will take place outside the domain of science. It will be metaphysical and thus in the realm of philosophy and, over time, mythology.

Perhaps we can even say that the *less* a final theory has to say of existential significance the *easier* it must be for a modern mythology to develop — there is an openness in such an emptiness. And beauty and poetry too, I might add; and poetry runs in the blood of any mythology.

I believe that our civilization is in desperate need of stories that speak to our hearts and not to our intellect. Stories, which are true because we *know* them to be true and where we do not need *analysis* and *surveys* to recognize what all societies before us have known and understood and cherished. Science is important, it has helped us and transformed our lives. It is an integral part of our intellectual lives, where the acquisition of knowledge and insights is a genuine force of nature — and a force of good. But science cannot satisfy our need for love and meaning and spiritual insights. It cannot and should not replace that ancient faculty called *wisdom*.

In an odd way, I believe that our search for scientific truth, our search for a final theory and the scientific project in its entirety, is ultimately a search for precisely such stories. It is an intellectual search for something that lies *outside* of the intellectual.

In the Book of Job we find the question:

> *"Where shall wisdom be found, and where is the place for understanding?"*

which is followed by

"Man knoweth not the price thereof;
neither is it found in the land of the living"

and

The depth saith, "It is not in me: and the
sea saith, it is not with me".

That's us, we have been searching for wisdom and understanding in the land of the living, in the depths and in the seas and everywhere else for a very long time. But like Job, we have not found the wisdom that we seek. Job found the fear of a wrathful God and we — who live in an era where God is long dead — will find ... what?

Where will humankind go when there is nothing left to conquer? If western man goes anywhere, I think he will turn inwards, to the third door, the subjective element in the universe. Consciousness.

The looking itself. Who is it that so desperately wants to know the truth? Who is the looker? When it is clear that answers cannot be found *in the land of the living* then I think that western man will turn around and direct his attention to the second element in the equation, namely consciousness itself.

Because there *is* a second element here. The objective always comes with a subjective; the Universe that we observe is precisely that: *observed*. We do not have access to, nor knowledge of, an unobserved universe.

Where else could they go? It's either that or resignation. It's either that or regression.

There are two ways to approach consciousness: from the inside and from the outside. The mystical approach and the scientific. East and west. Western man have already chosen the outside approach, the scientific. But gradually, perhaps as a parallel movement and as a reaction to the discovery of a final theory that is essentially empty, I imagine that western man with his hunger for truth, will eventually look inwards at the looking itself. And if that happens, then I think he will write a new story.

Then we will have a new mythology. If.

CHAPTER 24.

CROWDFUNDING

Water is ice
that has let go

— **J.M.G.**

It's the summer of 2015. I've spent the past months biking through the Alps in Austria and Slovenia, thinking about anything but physics.

And now I'm back in Denmark. I'm at the Ørslev Monastery in northern Jutland, a refuge for writers, artists and anyone who seeks a quiet place to work.

I've sold my flat in Copenhagen and am living on the money from the sale. I'm a nomad, a modern nomad with a computer and a backpack full of shirts and books. I'm in the middle of the river, trying to navigate across this beast.

This monastery is old. It dates back well before the Reformation, with beautiful white stone walls and high ceilings held by massive oak beams, with vaulted corridors, stone floors, and heavy wooden doors. The atmosphere here is calmer than calm. When I lie flat on the floor in my room, I sense a peacefulness that I have always known but never had.

I've decided to apply for research funding again. Since my last grant four years ago we have obtained good results, I'm confident of that. Our paper on quantum holonomy theory is in the review process and should soon be published. I feel that I have a solid platform for an application, my chances must be reasonable.

I spend a good deal of my time at the monastery working on grant applications. I divide my time between that and my work with Johannes.

However, a brief survey of the various Danish research foundations soon reveals that most of them do not accept applications from scientists who are not employed at a university. I find two exceptions to this rule: a major private foundation associated with a Danish brewery and the Danish Council for Independent Research — the largest provider of public funding for Danish research.

I first contact the private foundation to inquire about how I might best submit an application based on our recent results, which I briefly outline. I immediately get a reply from the director of the foundation himself, a highly regarded experimental physicist, who informs me that I am not eligible for funding and that he will not consider my application. After inquiring as to how this might be, he writes that I don't have the right profile — I have to be an internationally renowned professor with a *passionate* approach to research — and then he suggests that I instead seek a career in the private sector.

I then turn my attention to the second option and submit an application for a three-year research grant to the Danish Council for Independent Research. I arrange with the Niels Bohr Institute to provide housing for the project in case I get a grant. In my application I describe our research project and our recent results. I elaborate on the fact that Johannes and I have one of the very few research projects in the world that has a concrete strategy for a fundamental theory that involves unification and that we already have certain results.

Their review process is quite long but eventually, I get a reply — a rejection.

The council provides a brief explanation for their decision, in which they write that my application is in the third category out of three (not good) and that their decision is based on three main factors:

1. *That the project does not contribute sufficiently to the internationalization of Danish research, as the project is based on an already existing collaboration with a single Danish researcher located outside of Denmark.*

2. *That the project does not in a sufficient degree involve research education, as there has only been a request for funding for the applicant himself.*

3. *That the project does not in a sufficient degree document that the activities will benefit Danish research in the long term, as the applicant is not permanently associated with a Danish research institution and does not state his long-term career perspectives.*

I read the letter once more to see if I missed something.

There is not a single word about the content of my research in the letter.

The members of the council come from a wide range of disciplines, where half of those with expertise in theoretical physics work in string theory at the Niels Bohr Institute. This year the council handed out almost 9 million Danish kroner to researchers in string theory — of which more than 6 million went to scientists at the Niels Bohr Institute.

The sun is shining. The walls of the monastery are thick, more than a meter. They radiate a deep silence absorbed through centuries. I close my computer and blow out the candle on my desk. I get my jacket and go out.

As I walk out, I close the door behind me.

May 5th, 2016

I wake up because my phone is ringing. It's my brother,

- *"Hey, Jesper, have you checked your crowdfunding campaign today?"*

I look at my watch, it's half past eight in the morning,

- *"I think you should check it out, it's crazy".*

I don't have internet access at home, so I bike down to the public library.

I got the idea for a crowdfunding campaign from a radio program — some guy who didn't have the money to go to his friend's wedding crowdfunded the ticket — and immediately realized that it could work for me, too. But it was with mixed feelings that I began to set up a campaign for myself. As far as I knew, nobody in theoretical physics had done this before and certainly not with the title I had in mind — *A Theory of Everything* — and I worried how the scientific community would react, what it might do to my reputation. Would it hurt the standing of my work with Johannes? Also, I knew that such a campaign would automatically put me in the center of a good deal of media attention, something I was not comfortable with, and I knew that for the campaign to be successful I would have to seek out this attention myself. I spent a considerable amount of time contemplating how to do this, what my boundaries were.

But the upsides seemed so much greater than the downsides. With a crowdfunding campaign, I would no longer rely on research grants, which nobody seemed willing to give me anyway. I would completely circumvent the traditional system based on university research and above all, I would be able to maintain my freedom. I loved the idea.

But what rewards can you offer in a crowdfunding campaign that produces nothing but unintelligible formulas and technical research papers, which no one outside an exclusive circle of professional nerds understands? The only real idea I could think of was a book, which was very attractive to me as I've always liked writing — poetry, prose, *words* — but I've never dared to be serious about it. With a crowdfunding campaign, it would be a necessity.

When I arrive at the library and open my computer, I am overwhelmed by what I see. During the previous months,

I have mostly focused on the immediate tasks — build a network, make a video, write the campaign text, get colleagues to provide endorsements, figure out what tax rules apply to crowdfunding — and I never really contemplated what it would be like to go live with the campaign. I am generally on the introverted side and to now see the list of backers grow well beyond a hundred and continue climbing is really a bit unsettling. When I set up this campaign, I thought about what it would enable me to do but I never really considered that there would be real people on the other end — real people contributing real money. But the emails that now land in my inbox are very real indeed and although they are all positive and encouraging and full of goodwill, it is a bit like walking into a public swimming pool with no pants on (something I experienced as a kid, I forgot them in the locker room).

For several weeks I do the same thing every day: I bike down to the library, get a cup of tea and watch my crowdfunding campaign grow.

And as it grows a new sensation grows in my stomach. I start to believe that this can work. I see a path forward. My own.

June 16, 2017

It looks complicated.

The only light in the cabin comes from the dying fire in the fireplace and my computer screen.

The page is full of complicated formulas. They are beautiful. At first glance, they appear to be unintelligible scribbles. I find an elegance in that, it's beautiful — a beauty that fades with comprehension. I find beauty in the logic too, and that beauty expands with understanding and clarity, that is the beauty of perfection.

I place two pieces of wood on the fire and then scroll down in the document. Page after page with formulas. It's a long argument.

I go back to the top and begin to read. Deciphering hieroglyphs. Johannes sent me his proof and asked me to check it, to see if he has missed something. As if I am any good at that. I am a pirate and not an accountant but right now Johannes needs an accountant. So, I read the first sentence and resist the temptation to just jump ahead and skip the thinking.

It started rolling last summer. Around the time of the crowdfunding campaign, it set something in motion, or perhaps it was just a coincidence, but around the time of the campaign, Johannes got a couple of ideas.

Good ideas.

And then they kept coming, Johannes threw ideas at me, I worked on them, threw some back at him, got stuck, and then he threw some new ideas. And so on.

It felt as if we had been hammering on a wall in front of us not knowing whether it was the basic wall of a house or a mansion or palace or if it was the massive wall of a solid mountainside. And it felt like we had been hammering on this wall for a long time, a very long and exhausting time. Our backs were sore, our hands swollen, everything had begun to hurt, and a deep feeling of fatigue had started to grow in the dark corners of our hungry souls. But then Johannes had suddenly peeked around the corner, where he had found an entrance. A door.

And when he tried the handle it was open. The door was not even locked. That is what it felt like. He found a door and it was open.

And then we walked in.

The island of Møn, 15th of April 2020

I found it. This time I'm certain. I've checked my computations several times, I've slept on it and it still adds up. This time is different, I don't see any mistakes.

I began searching in early March. Johannes and I have been working throughout the winter on the construction of the Bott-Dirac operator but then he checked out on paternity leave. I had planned a vacation but when the Coronavirus hit the world I thought twice about my trip to Germany. Instead of a hike in the Harz mountains, I stayed at home and kept on thinking about the Bott-Dirac operator. There was something that bothered me. A technical detail. An itch.

And then I started digging. I was convinced that the solution to this problem exists. I could feel it. The Bott-Dirac operator has to satisfy several technical requirements and the more I dug myself into the details the more certain I became that a solution had to exist. But I couldn't see it.

It took me six weeks. Six lonely weeks where I didn't see anyone. I stayed in my cabin in the countryside and worked day and night on the problem. It became an obsession, an obsession squared. I wasn't going to let it go until I had solved it. My living room was full of papers, notes with formulas; I often woke up at night thinking that I had it and raced to my desk to write down my half-lucid thoughts, only to find minor misunderstandings and zero divine interventions the next morning.

But yesterday I ran out of mistakes and finally wrote down the solution.

I have gone out for a walk in the forest to get some fresh air and watch some trees. It's a good place to think.

I have the feeling that I am being guided by the mathematics. Every time we solve one problem another one turns up that leads us in a new direction. So far, all the problems we have encountered have had solutions.

I don't know where all this is leading us, but I feel as if we are being led. We just follow the trail laid out for us. Small white stones marking a path through the forest.

Hannover, August 2020

It is almost finished.

I can't wait any longer. I can't sleep either, my nights are filled with restless hours, when my mind, tired of running too fast for too long, cannot let go of my thoughts.

I'm sitting in the library at the Mathematical Institute at the Leibniz University in Hannover. This is where Johannes has a permanent teaching position and I have come here to finish our paper. We are almost there.

I think that our bodies too, know that we are close. They have been exerting themselves for a long time and now when the goal is near, they suddenly feel terribly tired and heavy. We have to hurry up and finish the paper before our energy runs out.

I begin to read the papers again. This is what we started searching for some 16 years ago. This is the mathematical construction that Johannes first outlined during his first visit to Iceland — we have pursued this idea for so many years and now we have found it. I *knew* it existed, the theory, I just knew it, we both did. We could smell it and now we have found it.

And it is so simple. It is much simpler than I expected. Almost too simple. It scares me — I would feel safer if it was complicated. Complicated means difficult, which in physics translates into accomplishment and respect from peers. Too simple opens the door to too stupid and too stupid sounds like unmet expectations, ridicule, and serious misunderstandings.

But simple is good too, right? There is grandeur in simplicity, and poetry, lots of poetry. All mountains are simple; they are large and point towards the sky. A complicated mountain is an artifact, a fake. At least that is what I try to tell myself.

It is beautiful too. It excites me. There is true beauty here, much more than I had ever hoped for. And it is so different from what I had expected. We were looking for a theory of quantum gravity but found something else. Something *better*, more elegant, more *natural*.

I feel both excited and nervous. We are so close, and I can't wait any longer. I need to rest but I can't rest until we have crawled across this finish line. Johannes is tired too, I can tell, we have even had a minor quarrel — we never quarrel! — moments of irritation, when my gentle nudging triggers his tired soul to lash out.

I look out the window and watch the busy road with streetcars and the afternoon traffic. I have sat in this library so many times during my countless visits, times when I have longed to be where I am now, reaching the end and finally seeing it. And now I just feel exhausted.

And yet, this is just a new beginning. It is written all over the pages that I am reading. This is not a finish line; it is not a conclusion but rather a prologue to a story yet to be written. With this paper, the work really begins. We have set up the theory, the walls and the roof, but we haven't looked inside to see what it really contains. We have peeked through the windows, but we haven't walked through all the halls and rooms and odd corridors. This is a task for the next decade.

I know in my bones that we are on the right track, but that knowledge cannot be published, it cannot pass through the peer review process. It needs to pass through our frontal lobes first. But these frontal lobes are getting tired.

I take a deep breath. The street outside is still busy. It always is. I turn around and continue reading.

✡

Sunday 23rd of August 2020

I uploaded our preprint to the archive yesterday, it will appear tomorrow. Our theory. Now it is out.

Tomorrow I will travel down south to visit my friend Hugo, who lives by the coast.

To relax. Finally. To sit by the sea and watch the waves and finally relax.

Something feels different today. I slept for 12 hours last night. And I didn't think about mathematics when I woke up.

I think that it has left me. The idea. That it is gone; let me be and moved on.

I walk out into the garden and sit down under the large birch tree. It is a different garden, but the tree is the same. There is no wind today, the tree is silent, perhaps it has nothing further to say. I am fine with silence, it feels like an old friend. We don't need to talk to feel comfortable in each other's company.

I sit for a while, thinking.

Who helped whom? Did I help this idea into the world —
or was it the idea, that helped me?

I found it, back then in Tibet many years ago, I took it in
and carried it through the years, nursing and nourishing
it. But it nourished me too, I can feel that now that it has
left me. The idea carried me too. The strength came from
it, not from me.

I lean back and watch the clouds. They too are silent. It is
all so silent today. As if there is nothing more to say — as
if there is nothing more for me to say.

EPILOGUE

I have come down to the beach to watch the waves and to collect seashells for my niece.

This beach is one of my favorite places. I love the monotonous sound of the waves as they break onto the fine sands and silently reach for the unknown beyond the dunes.

I walk barefoot along the beach as I watch seagulls surf the invisible wave above me. The sand is cold and wet. Sometimes I pass through a patch of fine stones, all rounded by their time spent in the sea. The water too is cold. I smell salt and seaweed as I watch the horizon, straight and infinite and beautiful out there beyond the ocean where it guards the home of the sun. It is patient, I imagine, knowing it has time on its side.

I stop to watch the waves. They roll silently towards the coast, regiments of straight lines, one after another, like shadow-horizons traveling towards the land of the living with tidings from the sun. They must have traveled far I imagine, days, weeks, to arrive here on my beach where they break with a sigh of relief. It is as if this deep ocean carries an eternal longing on its surface, a longing for what it is not, a yearning for an impossible union, a never-ending restlessness that moves forward, onwards, forever onwards, the tick-tock of time without time.

I turn around and continue my walk along the shore. I pass a large trunk of driftwood and a dead seagull caught in some rope. As I walk southwards, I watch the drifting clouds; they carry rain but it's not for me.

Once in a while, I pause to listen. Carefully. Listen.

I can almost hear it.

THANK YOU

This book is based on my long collaboration with the mathematician Johannes Aastrup — and I would like to thank him for his spirit of adventure, his friendship, and his intellectual openness on this long treasure hunt of ours. He has truly proven himself a worthy pirate.

I would also like to thank my friends Angela Warburton, Hugo Rodriguez and Jacob Højland, as well as my brothers Olav and Simon for their unwavering support, even at times when my ship was taking in water or seemed completely lost at sea.

And I would like to thank Hans-Jørgen Mogensen for constantly backing me and Birgitte Højland for her help with the book. I would also like to thank the physicist Christoph Stephan, Holger Merlitz, Mattias Thing and Paolo Bertozzini for reading my book and providing helpful feedback, both professionally and creatively. A special thanks also to Trevor Elkington, Stephan Mühlstrasser and Nick Taddeo for test-reading my book and providing many corrections and insightful suggestions. Also thanks to Eliane Pohl for her help on completing the English edition and to Britt Tippins for our teamwork on the Danish edition. Finally, I would like to direct a very special thank you to the designer Sten Jauer, who in the eleventh hour fell out of the sky and created the beautiful cover for this English edition.

And then there is the 2016 crowdfunding campaign, where hundreds of people threw their trust and their money at me with no other guarantee than my word. Their trust and enthusiasm gave me a much-needed injection of instant motivation; it was extremely encouraging and helpful — not only financially but equally importantly morally — to experience their support for my campaign. I am deeply grateful to all of these lovely strangers, who answered my call for help. My book is dedicated to you!

On one of the first days in my crowdfunding campaign, I received an email from the British philanthrope and technologist Ilyas Khan, who wrote that he had just transferred 10.000 dollars to my campaign. It is a great pleasure to thank him for this very generous contribution, I am honored by his faith in my work and I hope that I have been able to meet his expectations.

I am also greatly indebted to the following list of backers in my crowdfunding campaign: Ria Blanken, Niels Peter Dahl, Simon Kitson, Rita and Hans-Jørgen Mogensen, Tero Pulkkinen and Christopher Skak. On my homepage, I have a full list of the more than 400 people who backed my crowdfunding campaign.

I am also greatly indebted to my two largest sponsors Regnestuen Haukohl & Køppen and the entrepreneur Kasper Bloch Gevaldig as well as all my other sponsors and donors, whose names are too many to mention here but whose support I deeply appreciate.

Finally, a special thank you to Heidi.

Made in the USA
Middletown, DE
28 May 2021

would be interested in another try. He said yes, then I asked if would consider doing a little "manscaping" before next time. He surprised me by asking if that was something I would want to help him with. I said: "I would." And so it began.

He had not been with a man before me. And I'll admit, at first he was not 100 percent overzealous. But he was at least curious, and teachable. It took a couple of times to find our rhythm and learn what each other enjoyed. I had never been with anyone sweeter, more gentle or willing to please. One time, he asked me to tell him one of my fantasies, and I did. The next time we were together, he had come prepared to create that fantasy for me. I loved being with him, whether it was sexual or non-sexual.

We were not out in public very much, preferring to stay in and enjoy dinner at home. But in 2002, as my 50th birthday approached, a friend offered to have a dinner party for me. It was a small, intimate evening and I wanted Mr. 4-Names to be by my side. He was hesitant at first but knew how much it meant to me. He had met my hostess and another friend. I included my trainer, and for some reason, Mr. 4-Names did not mind that he knew something was going on between us. A couple of other friends, including my college girlfriend Carol, were there — and I enjoyed their company.

When we first came together, Mr. 4-Names had a motorcycle and not too long afterward had an accident, injuring his leg. I wanted to take care of him and asked him make out a grocery list for me. He was so strict with his diet that I wanted to make sure I bought the right things. He tried to talk me out of it but finally relented. I felt so happy being able to care for him. I also encouraged him to sell his bike, which he did.

LOVER TONIGHT

I was in love, so long ago, you know, it's hard to say just when,
I've blocked it out, because of the pain, and said, I'd never, ever love again.
Well, several years have come and gone, and old feelings, just seemed to die.
Life alone, well, it's not so bad, and cold hearts, don't ever have to cry.

So what about you? Just look what you've done.
You've captured my love and got my heart on the run.
I know I said, never again, no more lovers, only friends.

And now I find that I'm falling in love.
But it's just friendship, you're thinking of.
Well, I've got friends on my left and my right,
What I need now, is a friend, who'll be a lover tonight.

Mr. 4-Names became certified as a personal trainer and our schedules became more complicated. We decided to set aside days and times that we could be together, and as strange as it may seem to others, it worked well for us. I had no illusion of him busting out of the closet door, but I truly felt he cared for me and wanted to make time for me. He would still date women on occasion and eventually, one became a bit more serious. However, he still would make our scheduled times. I did not want a full-time relationship, so this arrangement was perfect. And for four years, we were together in our own little world.

One night, as we sat down for dinner, my phone rang. I still regret ever answering it, as it turned out to be his current girlfriend. She needed to speak with Mr. 4-Names. He had told her that we were working out, but she had been by the gym and his house. Fortunately, she did not know where I lived. I put him on the phone and when he hung up, he said he had to leave but that he would be back later.

True to his word, he did return, but not in the way either of us had planned.

He called me about an hour later and said: "We're coming over." What? I had no idea what that meant, but I found out soon enough. They rang the bell, and when I opened my door, she charged in and immediately began to yell at me. She had wanted to know what was going on, and Mr. 4-Names had confessed and told her the truth.

Unfortunately, the reason she was tracking him down was because she had just found out that she was pregnant. I may have been there first and may have had him longer, but she played the trump card and did the one thing that I could not: She roped him into marriage with a baby.

She ranted and raved and called me all sorts of names. In her mind, I was sick and had seduced him — everything was my fault. Finally I looked at her and said: "Get out." I wanted to say so much more but as I looked at the man I loved and cared about, I saw the pain in his eyes, and I was not going to add to it. She continued to throw her tantrum and again, I said: "Get out of my house. NOW!" He looked at her and said, "Go. I'll be right out." When she hesitated, he said again: "Go, please."

She stormed out of my home, slamming the door. He looked at me and said softly: "I'm so sorry." And I said: "I know. I am, too." I hugged him, gave him a quick kiss, and told him that I loved him and if ever he needed me, all he had to do was call. As he closed my front door, I thought of a line from the song that I wrote for my father after my mother passed away: *Silence echoes through our home.*

He did the right thing, or what he felt obligated to do. He married her, moved north so she could be near her family, and even fathered a second child. Although they are no longer together, he continues to live up north and be a good father to his teenage children. We have reconnected, if only to send the occasional message. But I am grateful that he is still in my life. I have not dated anyone since my 4-Name Prince Charming.

YOU MAKE ME CRAZY

You, make me crazy, when I'm with you, or without you,
What will I do, I don't know, I've lost control, of my heart and soul,
And you, you seem just fine, while I'm losing my mind,
In love with you.

You make me crazy, when you say those things, that make my heart sing,
Oh what joy you bring, everyday, when you say, I love the way,
That you make me feel, our love is for real,
How I need you.

Well, I don't know, what you see in me,
But I'm just grateful that you do.
Our love must be built on insanity,
'Cause, I'm just crazy in love with you.
Yes, I'm just crazy in love with you.

After Mr. 4-Names moved away, I sold my art gallery and worked three years for the new owner before being let go. The economy made it difficult to find another job as I pursued leads in my fields of interest — I found no one was hiring. Lonely and feeling untethered, I began making regular trips to central Florida to visit my dad's two sisters. I toyed with the idea of moving there but could never bring myself to follow through.

Several artists were disappointed when I sold my previous gallery. They did not feel a connection with the new owner and had asked that I continue to work with them. No longer having a gallery, I began to rent space inside of other businesses and display their artwork, letting the different businesses handle the sales. Eventually I had four locations in Nashville and one in Memphis. Realizing how much I missed my art gallery and with the noncompete clause no longer in effect, I decided to open a new gallery — York & Friends Fine Art. I found a wonderful midcentury building for lease less than a mile from my childhood home. I started small but have been able to expand as the business grows.

I had no interest in another relationship, and yet someone from my past life came along that provided a distraction now and then.

I met "Red Truck" more than 20 years ago — he was divorced at the time and wanted nothing more than a quick, no-strings-attached rendezvous, and I was more than happy to oblige. Drifting in and out of my life over the years, he reappeared during the time I had our family home emptied and on the market. Although he had remarried, he still enjoyed a little something on the side.

Long after Mr. 4-Names had moved on with his life, "Red-Truck" stopped by my new gallery. We had a brief encounter, and he continued to drop in regularly. I enjoyed being with him and invited him to my home. In spite of the obvious reason for getting together, we had developed a friendship that led to conversation. Plus, we both were getting older and long past the raging hormones. I enjoyed his company, which after so many years, felt relaxed, familiar and comfortable.

I had one "rule" regarding our relationship: that he was to not drop by my home without calling. Working in retail, I spend a great deal of time with people — and my home is my refuge. I'm happy to have guests but want it on my schedule. All he had to do was call me and we would make a date. There were times he would call me to make sure I was at the gallery, so I did not think it was too much to ask for him to call before stopping by my home.

Although I had his cell number and could easily look up his work number, I would never call him. He had asked me not to, and I honored his request. He did not need his wife, or co-workers, questioning a call. Plus, because of the location of my gallery on a busy street, he would often park so his "Red Truck" would not be seen.

For some reason, he kept stopping by my home, even though I continued to ask him not to. It reached a point that I would not answer my door. I began to keep my blinds closed and if watching TV, I'd keep the sound down. I had a garage without windows, so he could never be quite sure if my car was there. I would hear the rumble of his truck and the door of the cab close right before my doorbell rang. I would sit quietly until I heard him leave. When this happened four nights in a row, I became concerned. There was desperation in his behavior, something I knew from my own experience.

My best friend was aware of "Red-Truck," but it was then that I gave her his actual name and where he worked, just in case. "Just in case?" she asked. He had not given me any reason to be fearful of him in the past, but for some reason the nightly drives by my home gave me pause. Finally, he stopped by the gallery and as much as I hated to end it, I did. I said: "This isn't working for me any longer." I continued with: "I asked only one thing of you, and yet you continued to ignore it by stopping by my house." I did not want to admit that I had been inside when he stopped, but I did tell him that my neighbor had mentioned someone in a red truck circling the neighborhood.

I don't think I was ever in danger. In fact, I'm sure of that now. I'm confident I was projecting my urgent and reckless past behavior on his actions. Still, I had reached a point of feeling uncomfortable and decided it was best to walk away. He has respected my wishes.

I catch a glimpse of him, now and then.
You know the mind plays tricks, you just can't win.
You think you're safe, no more pain,
Then you close your eyes, and it starts again.

ELEVEN

FLASHBACKS & MEMORIES

On May 16, 2017, I posted this message on my personal Facebook page.

"I may have just overstepped, but I don't regret it.

After a morning appointment, I stopped by the McDonalds adjacent to Centennial Park. I sat in a booth next to a young couple with two small boys, and heard the parents tell their young boys that after breakfast, they were going back to the hotel. This led me to assume that they were from out of town.

The patrons in this particular restaurant are always an interesting mix. Being situated next to a public park, it attracts tourists, which I believed this family to be. It also draws from the neighboring offices and retail stores, plus those that may be homeless and look as if they slept the previous night, under the stars. There was one man fitting that description who was in my direct line of sight. I noticed he kept staring at the family seated next to me. In fact, his gaze was so intense that he never noticed me watching him.

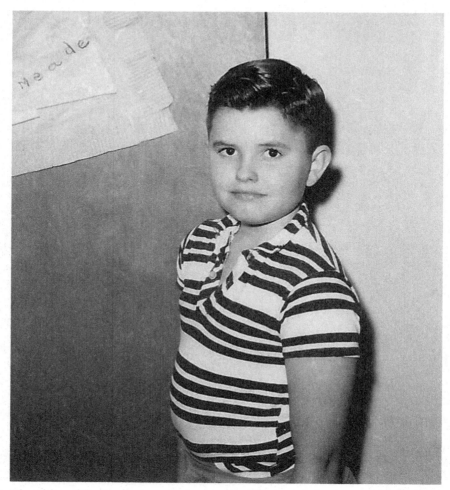

The older of the two boys, stood up and said he needed to go to the restroom. The father looked past the counter, saw the restrooms and pointed the way. The man who had been watching so intently got up and followed the boy into the restroom. Sitting in the next booth, I was facing the father and got his attention and told him that he needed to go with his son. He looked at me, as if to say mind your own business. I said much more sternly: 'Go check on your son, now.' His wife looked at me, then at him and said: 'Go.'

When he and his son came back to the table, he thanked me. I don't know for sure if the child was at risk, but I was not going to take the chance."

I had maintained my composure until I got to my car. Then I fell apart. I began crying hysterically. I realized memories of my own childhood still haunted me. Once I pulled myself together, I drove on to work. I decided to share this story with my Facebook friends, not to receive a pat on my back but to make them aware of possible dangers out there and to tell them not be afraid to speak up.

That post received nearly 500 "likes" and over 170 positive comments — more than any previous post that I have made. There was overwhelming support that it was the right thing to do. We need to watch out for each other and to speak up if we see something that might be questionable. I appreciated the outpouring of love and am glad I was there that morning.

Not only do we need to watch for signs of abuse, it is also important to be there when someone needs to share their experience. It's hard enough, just finding your voice. My heart broke for those who told me that because of who their abuser was, their parent did not believe them. Surely their uncle, or friend or boss would not have done that, would often be the response. They simply must have misunderstood. How painful it must be to find the strength to talk about your abuse and not be believed.

A woman came into my gallery for the first time, on a mission to find a special work of art. Her husband was out of town, but she said she'd return with him in a few days. True to her word, she and her husband stopped by a couple of days later. They both found several things they liked but just could not agree on that one perfect piece. They did purchase one of my books, however.

A few days later, they were back again. And as they came through the door, the husband announced he had just finished reading *Kept in the Dark*. He commented about the vintage photos and the letters at the heart of the story. He even mentioned his birthday is the day after mine. They started to look around, but I could tell he still had something to say. Eventually, he mentioned that his late brother and I were in college during the same years. I studied at Belmont University while his brother was down the street, attending Lipscomb University. And then, in a soft voice, he confided that his brother had also been molested.

It wasn't the time or place to ask questions. He said what he came to say and then thanked me for writing my book.

He is just one of many who have reached out to me since the book's release. Driving is often the one chance I am alone with my thoughts, and I find myself thinking about the things people have shared with me — and my heart breaks. I have shared that, if you happen to stop alongside me in traffic and look over and see tears streaming down my face, please know that I am okay.

Friends have told me their personal experiences — things I never knew or suspected. Like me, they have been good at keeping the past hidden. One of my childhood friends sent me the most heartfelt message after reading my book. She said she always saw me as the child with the smile and had never realized until now what that smile had been hiding. She also shared that one of her children had been through something similar. I told her that I felt no child or parent should ever have to deal with this.

Friends and strangers have told me of their pain, or that of their child, their spouse, or their friend. It may have been a one-time encounter, or something that went on for much longer. The violator so often was a family member or someone known and trusted. There were even a few, like me, with multiple abusers over multiple years. In my case, it reached a point at which I simply came to expect it.

Regardless of the number of times or the number of perpetrators — one time is one time too many. Somehow, we need to find a way to protect the innocent. I decided one thing that I can do for them is to listen and to be strong — but that doesn't mean there might not be a time when tears will be the only words I can say.

I had a television interview when *Kept in the Dark* was released. Questions and answers from my PR firm had been forwarded in advance. I had practiced answering the pre-written questions and felt prepared. I arrived on time and met the host of the show. She then passed me along to her co-host for the interview. I asked if he had the list of questions. No, he had not been given them, and we were just minutes from going live. He quickly handed me a yellow legal pad and said to jot down questions. My mind went blank. I scribbled a couple of things and then it was time. He began asking about my father and the abuse charges that had been brought against him over 50 years ago. I answered him and then added that I had been abused, not by my father, but by multiple men through the years. He asked a couple of follow-up questions, and then we were done.

I tried to hide my disappointment from not being asked what I had prepared myself for. I thanked him, and he then offered to walk me out. He was in the middle of a live program and yet he wanted to walk me down the long hallway back to the front lobby. As we made our way, he told me that I was brave to come forward. He also asked to keep the copy of my book and added that he had been molested many times as a child. I left the station thinking the interview did not go as I had planned, but it went as it was intended.

I sent him an email when I got home, thanking him again and reminding him that I'd be back in town the following week for my book signing. I also offered to listen, if he ever wanted to talk. He said he was not ready now, but he thanked me and said he would keep me in mind when the time came.

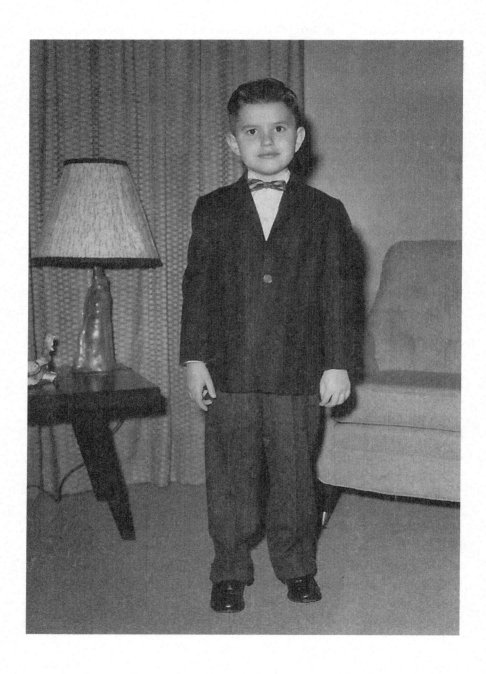

My emotions have been on a rollercoaster ride. The overwhelming and hysterical feeling that engulfed me after seeing the movie *Spotlight* became the catalyst to finally go through the box my parents left for me. But up to the complete meltdown that happened after the McDonald's incident, I thought I had my feelings buried. After *Kept in the Dark* was released, friends as well as strangers reached out to me and shared their own personal stories.

Post-traumatic stress disorder (or PTSD) is often described as a mental condition resulting in reactions, whether emotional or physical, after experiencing a traumatic event. I've been told I have experienced PTSD. Does that excuse or explain my behavior? It is difficult for me to say yes. There are so many people who have experienced sexual abuse, and reactions can vary.

There are many symptoms of PTSD, some of which I don't believe I ever experienced. There are others that have my name written all over them. I don't remember any physical pain, nightmares or flashbacks. But depression has knocked on my door many times through the years. Bedwetting in children can be a sign. My mother's letters mentioned me often not getting through the night without wetting the bed. This is something that continued, off and on, for several years. Also, returning to regressive behaviors, such as thumb-sucking, something I did until college.

Then there is the symptom of risky behavior. As I have said, the warning lights never went off in my head when I stepped over the line and found myself in a downward spiral, during which suicide seemed like the only solution. I saw no way out and no way to change who I was so that I could be what I perceived as "normal," like the other kids.

And was I thrill-seeking when I took that young man into my church Sunday School classroom at the age of 16? Or when I took the sailor home to shower? Or, as an adult, when I found myself circling the downtown streets of Nashville to pick up a hustler so that I could film him?

Another symptom is repression. Did I intentionally block some of my childhood memories until now? I never would let myself, until recently, think about or linger on those memories of my youth and the downtown movie houses and men's rooms. Emotional numbing is something often talked about with PTSD, during which you learn to not feel anything and sometimes withdraw from social circles. Yes, check that box. There were times I felt that way.

Guilt and shame are also symptoms and were very much a part of my teenage years, when I knew I was different and yet felt unable to conform to what I thought was expected.

I truly believe we are born gay. My childhood antics, described in my mother's letters when I was 2 years old, indicate to me that I was headed in that direction. Because of that, I have often wondered if being seduced by a man made my molestation less horrifying to me? What if it had been a woman, or women, who had molested me sexually — would that have been more traumatic? I don't know the answer, but there were times when women did make me uncomfortable, when they took the approach that the only reason I thought of myself as gay was because I had not found the right woman. In fact, there was a woman from one of my church job experiences who practically tried to rape me, in my own home, determined that she could set me straight.

The symptom that resonates the most with me is the difficulty in maintaining personal relationships. While I have been fortunate to maintain several long-lasting friendships, I have not had the same success romantically. In fact, my longest romantic relationship was for four years with Mr. 4-Names, and that was never a traditional relationship. It was scheduled weekly appointments, not really all that different from hiring an escort. I did care for him and loved him, in my own desperate way.

> *But tonight I'm so lonely, and I want to be held.*
> *I don't think I, can make it alone.*
> *So, if you think it's alright, if you think it's okay,*
> *Would you please take me home?*
>
> *And we'll just close our eyes, and fantasize,*
> *About a love for you and me.*
> *And somewhere in the night,*
> *Before the morning light,*
> *Maybe, we'll find our fantasy.*

I'm sure you have picked up on the number of times I have used the word *fantasy* in my writing, whether as part of the story or in one of my song lyrics. When I realized the repetition, my first thought was to find substitute words. But then, I thought, *there is a reason*, a purpose for the use of that word, beginning with my first writings in junior high.

I have often let myself live in a fantasy world because my real world has been painful at times. I have felt that I don't fit in — that I'm sick, grotesque, broken. I'll admit to letting myself dream, especially as a child — wishing my feelings were thought of as normal and accepted. And yes, through the years I have let myself project a facade, an invention of what I thought I was expected to be.

I keep asking myself, how does my story end? Now that I've faced my demons and my past — have I come to terms with it and am I able to move on? Have I made peace? Have I actually stepped out of the dark? Do

I blame the men that took advantage of me? Do I blame the church, or the movie theaters that unknowingly allowed young boys to be molested? Or do I blame my parents?

In writing this book, I finally sought help. At one of my sessions, I was told to imagine my mother sitting across from me. "What would you say to her?" and, "Don't you want to ask her, why?" My response was, "No." "Why are you protecting her?" As tears began to fall, I explained that now that I have learned what she endured through my father's arrest, her family and friends knowing, her nephew being the victim, having to leave her home of Miami and start over in Tennessee, with a young child and a husband she loved so much, that she was willing to forgive his sins… "How could I now add to her heavy load?"

Asking the same question, but with my father seated across from me: "What would you say to him?" It would be easier for me to speak to him — to ask him why, now knowing what he was accused of, why did he not try harder to protect me. If there was ever any doubt of my homosexuality, surely he already knew when he forbade me to play with another boy in elementary school. And if not then, what about when he caught me at 14 years old in a lie, then found me in a public restroom notorious for sexual activity? Wasn't that enough for him to see that I was headed down a road familiar to him, and that he needed to warn me — to stop me somehow? And if after all of that, he still had confusion, would it not have been cleared up by my suicide attempt?

I would want to ask him: Was it just easier to ignore all of the signs and cries for help than to admit his son had a problem and needed protection? Was our family that dysfunctional, that much in denial?

Of course, blame could be placed on society and the time period of the 1950s and 1960s. If sexual things were known, they may not have been talked about. Having same-sex feelings was thought of as wrong through my school years.

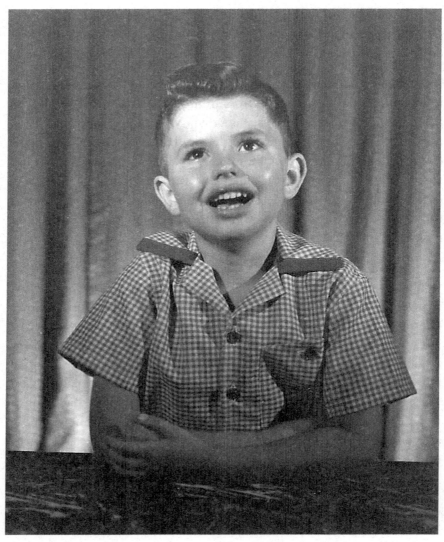

The church might preach love and forgiveness, but there was never a time that I thought the church would love and forgive me for what I felt. Then again, it was the church where I was first taught about expressing same-sex feelings in that children's Primary Sunday School classroom, with my pants around my ankles.

Once my first book was published, I received a message from a childhood friend I knew through church. She remembered her parents, along with others, going to our pastor with concerns regarding two men within the church family. Her parents warned her, along with her brothers, to report to them if they spotted the men. I was not given this warning but have to assume my parents might have been made aware of the possibility. I also realize this was just a short period after our problems in Miami and can understand if my parents wished to distance themselves from the controversy.

She could not remember the name of one of the men but had a couple of details she thought might be helpful. The second man who molested me is a vague memory for me, too. I don't remember his name, or think it happened more than a couple of times. She had been told he was asked to leave the church.

The other man she had been warned about was the first of my three church-related abusers. She did not want to give me his name, in case she was wrong. Instead she asked if his initials meant anything to me. I said yes and said his name, which she confirmed. She had been worried about saying anything and had hoped it would not upset me. Actually, the news made me feel a bit of relief. I had kept this information to myself since childhood, so there were times I feared I might have confused him with someone else. I wonder now if the two men knew of each other's proclivities and if they ever discussed me. Or if the first man passed along my name to the other, thereby putting me on his radar.

She contacted a mutual friend, who was part of the church family for as long, if not longer, as the two of us, and asked him if he was aware of any of this. Sadly, he was and said he knew there were other boys who had been molested. When she asked if I'd like for her to try and find out their names, I said no. I would never want to "out" someone or make them relive an experience that may have been painful. I have recently been open about my past, and if someone learns of that, has been in a similar situation and wants to contact me, then I am here for them.

I have asked myself (as well as sought advice from professionals) as to why I did not stop this impulsiveness — this self-destructive behavior, when I became older. Had my desperate need for attention become an addiction? Had the few moments of sexual pleasure been worth the hours of waiting, searching and risk-taking it took to find them?

The answer given has often been that I did not have a foundation in my childhood that allowed me to learn the difference between right and wrong. What I learned was picked up through secret rendezvous in my church and the downtown venues.

216 |

The American Psychiatric Association describes addiction as follows:

A complex condition, a brain disease that is manifested by compulsive substance use despite harmful consequence...Over time people with addiction build up a tolerance, meaning they need larger amounts to feel the effects...People with addictive disorders may be aware of their problem, but be unable to stop it even if they want to.

As much as I hate to admit it, this description explains to me what I have felt. "Compulsive ... despite harmful consequence." Needing a bigger fix each time to be able to feel it. And even though I was aware of it being a problem, there were times I did not think I would be able to stop. Does an alcoholic blame the person who gave them their first drink? Does the drug user place blame on their dealer? Where do I place the blame? I don't.

I believe my parents did the best that they could. Yes, I felt they did not hear my cries for help. But having my situation be so similar to what they had encountered in Miami, maybe they just could not deal with it or did not know how to handle what I was going through. Plus, I never told them I had been molested. We did not have the educational tools to warn us that we have today — so my parents might not have been knowledgable about the warning signs my behavior expressed.

We all have issues in our lives to be dealt with, by whatever means possible. I chose to create a smiling facade, in hopes that the real me, my true self, could stay hidden. I realize now that my mother probably did the same, in an effort to hide the ordeals our family dealt with in Miami. Once I learned of my father's crime, I began to feel my own facade crack. This allowed me to reveal a portion of my life that had been buried deep, which in turn was enough to have others open up to me with similar experiences. The walls I had carefully built around me finally came crashing down.

TWELVE

THE SOLID ROCK

I know now that my childhood foundation was built on sand and thus would often crumble beneath my feet. I am hopeful that it is not too late for me to rebuild my life on solid rock. Regardless of the experiences I was subjected to within the church, my Southern Baptist roots are still important to me — as are Edward Mote's words, based on Matthew 7:24-27, that became the anthem, *The Solid Rock*.

My hope is built on nothing less
Than Jesus' blood and righteousness;
I dare not trust the sweetest frame,
But wholly lean on Jesus' name.

When darkness veils his lovely face,
I rest on his unchanging grace;
In every high and stormy gale,
My anchor holds within the veil.

His oath, his covenant, his blood
Support me in the whelming flood;
When all around my soul gives way,
He then is all my hope and stay.

When he shall come with trumpet sound,
Oh, may I then in him be found;
Dressed in his righteousness alone,
Faultless to stand before the throne.

Refrain:
On Christ, the solid Rock, I stand;
All other ground is sinking sand,
All other ground is sinking sand.

Finally, I have unburdened myself. Like those actors I talked about earlier who removed their clothing because it was integral to the story, I have bared my soul. I don't know what the response may be. But to help me stay strong, I keep reminding myself of this quote from the late Elisabeth Kubler-Ross, a Swiss-American psychiatrist, who said: "The opinion which other people have of you is their problem, not yours."

Regardless of what anyone might think, and in spite of the fact that I have written this book, I feel that I am no longer desperate for attention. I know I have led a colorful yet flawed life, one that has been filled with love, excitement, loneliness and often despair. I have taken risks, unnecessary risks, but I have survived. I believe my life has purpose. And like my father before me, I am trying to make amends.

WHO WILL REMEMBER?

Lookin' through old photographs,
I remember the times we laughed,
And cried, so long ago,
They were good and bad, happy and sad.

Sitting alone, thinking of you,
Remembering family and friends we knew.
One thought keeps flashing through my mind,
What memories of our life, do we leave behind?

Forgotten memories, when we were young.
Locked in our hearts, they need to be sung.
But with no one to share the past,
Memories will fade, dreams just won't last.

Who will remember, after we've gone?
Who will remember, who'll sing our song?
Good times and memories won't be kept alive,
If no one remembers, the memory will die.

Lookin' through old photographs,
I remember the times we laughed,
And cried...

EPILOGUE

For nothing is hidden, that will not be revealed; nor anything secret, that will not be known and come to light. Luke 8:17

On a warm Sunday morning in December, I dressed appropriately and left the suburbs for downtown. I pulled into the crowded lot at First Baptist Church and was grateful to find a parking space. Arriving earlier than I intended, I sat in my car and said a prayer. As I waited, I reflected that the spot where I was parked had been the location of residential buildings when my family first joined First Baptist Church nearly 60 years ago. In fact, one building was always rumored to have been a brothel. I remember as a child thinking it seemed odd that it was next door to our church.

I had not been inside of the church building since my father's funeral 17 years earlier and had not attended regularly for a few years before that. Having just completed the manuscript for *Songs from an Imperfect Life,*

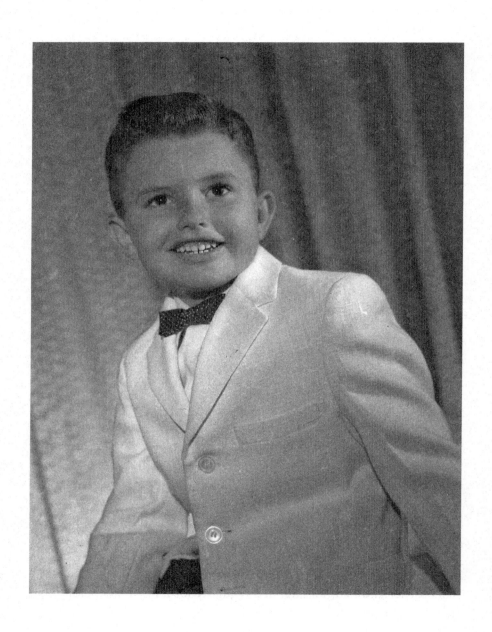

I felt a need to walk those halls one more time, as that church building held many of my childhood secrets.

A couple of my friends had offered to go with me, but I chose to go alone. I walked across the parking lot and entered the lower-level back entrance. My vivid memory of what had happened there was quickly diminished when I found a totally new entrance. I climbed the stairs, expecting to see a linoleum tile hallway on the third floor, but everything had been carpeted. I glanced into the open doors of what had been the Intermediate department only to find it had been remodeled and combined with the adjoining space. It did not look anything like I remembered from my teenage years when my father served as the department superintendent. I even checked the men's room on that floor, but found it too had undergone a total renovation.

Choosing a different route out of the building, I tried to blend in with the unfamiliar faces flooding the hallways as Sunday School ended and parishioners rushed toward the sanctuary for the morning service. Once I was back in my car, I took a deep breath and felt relieved. Those childhood memories no longer existed in that structure for me. Though I fear some memories will forever haunt my dreams.

CPSIA information can be obtained
at www.ICGtesting.com
Printed in the USA
LVOW13*2316110218
566163LV00001B/1/P